Angiogenesis and Anti-Angiogenesis in Hematological Malignancies

AF617956

Domenico Ribatti

Angiogenesis and Anti-Angiogenesis in Hematological Malignancies

Domenico Ribatti
Department of Basic Medical Sciences
Neurosciences and Sensory Organs
University of Bari Medical School
Bari
Italy

ISBN 978-94-024-0305-3 ISBN 978-94-017-8035-3 (eBook)
DOI 10.1007/978-94-017-8035-3
Springer Dordrecht Heidelberg New York London

© Springer Science+Business Media Dordrecht 2014
Softcover reprint of the hardcover 1st edition 2014
This work is subject to copyright. All rights are reserved by the Publisher, whether the whole or part of the material is concerned, specifically the rights of translation, reprinting, reuse of illustrations, recitation, broadcasting, reproduction on microfilms or in any other physical way, and transmission or information storage and retrieval, electronic adaptation, computer software, or by similar or dissimilar methodology now known or hereafter developed. Exempted from this legal reservation are brief excerpts in connection with reviews or scholarly analysis or material supplied specifically for the purpose of being entered and executed on a computer system, for exclusive use by the purchaser of the work. Duplication of this publication or parts thereof is permitted only under the provisions of the Copyright Law of the Publisher's location, in its current version, and permission for use must always be obtained from Springer. Permissions for use may be obtained through RightsLink at the Copyright Clearance Center. Violations are liable to prosecution under the respective Copyright Law.
The use of general descriptive names, registered names, trademarks, service marks, etc. in this publication does not imply, even in the absence of a specific statement, that such names are exempt from the relevant protective laws and regulations and therefore free for general use.
While the advice and information in this book are believed to be true and accurate at the date of publication, neither the authors nor the editors nor the publisher can accept any legal responsibility for any errors or omissions that may be made. The publisher makes no warranty, express or implied, with respect to the material contained herein.

Printed on acid-free paper

Springer is part of Springer Science+Business Media (www.springer.com)

Acknowledgments

This work was supported by European Union Seventh Framework Programme (FPT7/2007–2013) under grant agreement n 278570 to DR.

Contents

List of Abbreviations

ALL	acute lymphoblastic leukemia
AML	acute myeloid leukemia
Ang	angiopoietin
AQP4	aquaporin-4
ATL	adult T cell leukemia
ATP	adenosine triphosphate
CAM	chorioallantoic membrane
CBC	complete blood count
CECs	circulating endothelial cells
CLL	chronic lymphocytic leukemia
CML	chronic myeloid leukemia
DLBCL	diffuse large B-cell lymphomas
DVT	deep vein thrombosis
ECM	extracellular matrix
ECOG	Eastern Cooperative Oncology Group
EGFR	Epidermal growth factor receptor
EMEA	European Agency for the Evaluation of Medicinal Products
EPC	endothelial precursor cell
ERKs	extracellular-signal regulated kinases
ETS	expressed sequence tags
FACS	fluorescent activating cell sorter
FDA	Food and Drug Administration
FGF-2	fibroblast growth factor-2
FKHR	forkhead transcription factor
FL	follicular lymphoma
G-CSF	granulocyte-colony stimulating factor
GF	growth fraction
GFAP	glial fibrillary acid protein
GIST	gastrointestinal stromal tumor
GM-CSF	granulocyte macrophage-colony stimulating factor
HDAC	histone deacetylase
HGF/SF	Hepatocyte growth factor/scatter factor

HIF	hypoxia-inducible factor
HL	Hodgkin lymphomas
HOX	homeobox
HUVEC	human umbilical vein endothelial cell
ICAM-1	Intercellular Adhesion Molecule 1
IFN	interferon
IGF-1	insulin-like growth factor-1
IGH	immunoglobulin heavy chain
IgVhH	immunoglobulin variable gene segments
IKK	I kappa B kinase
IL	interleukin
IMIDs	immunmodulatory drugs
I-TAC	interferon inducible T-cell alpha chemoattractant
JAK	Janus kinase
LFA-1	lymphocyte function associated antigen-1
LI	labeling index
MALT	mucosa associated lymphoid tissue
MAPK	mitogen-activated protein kinase
MCL	mantle cell lymphoma
MCP-1	monocyte chemotactic protein-1
MDS	myelodysplastic syndrome
MEK	mitogen-induced extracellular kinase
MGUS	monoclonal gammopathy of undetermined significance
MiRNA	micro RNA
MM	multiple myeloma
MMP	matrix metalloproteinase
MVD	microvascular density
NF-kB	nuclear factor kappa B
NGF	nerve growth factor
NHL	non Hodgkin lymphomas
NOS	nitric oxide synthase
NSCLC	non-small-cell lung cancer
PCNSL	primary central nervous system lymphoma
PDGF	platelet derived growth factor
PDGFR	platelet-derived growth factor receptor
PECAM	platelet endothelial cell adhesion molecule
PI3K	phosphatidylinositol-3 kinase
PKC	protein kinase C
PTCL	peripheral T cell lymphoma
RTK	receptor tyrosine kinase
RT-PCR	reverse transcriptase- polymerase chain reaction
SCF	stem cell factor
SCID	severe combined immunodeficiency
SDF-1α	stromal cell derived factor 1 alpha
SDS-PAGE	sodium dodecyl sulphate polyAcrylamide gel electrophoresis

SelCiDs	selected cytokine inhibitory drugs
SLL	small lymphocytic leukemia
STAT	signal transducers and activator of transcription
TGF-β	transforming growth factor-β
TIMP	tissue inhibitor of matrix metalloproteinase
TNF-α	tumor necrosis factor alpha
uPA	urokinase plasminogen activator
VCAM-1	vascular cell adhesion molecule-1
VDAs	vascular disrupting agents
VE	vascular endothelial
VEGF	vascular endothelial growth factor
VEGFR	vascular endothelial growth factor receptor
VLA-4	very late antigen-4
WHO	World Health Organization

Chapter 1
Introduction

1.1 Angiogenesis

Angiogenesis (new vessel formation) occurs during embryo development and in postnatal life, cyclically in the female genital system and in wound repair. In these situations, it is limited in time and the result of an equilibrium between the activator and the inhibitor systems that together keep the microcirculation in a quiescent state, with very low proliferation and turnover of the endothelial cells.

The quiescent endothelium rests on a specialized form of the extracellular matrix (ECM)-the basement membrane-whose main constituents are laminin and type IV collagen (Ingber and Folkman 1989). Irrespective of the nature of the inducing stimulus, angiogenesis develops through five steps: (a) Basement membrane degradation by the proteolytic enzymes (metalloproteinases, collagenases, heparinase and plasminogen activators secreted by the endothelial cells, resulting in the formation of tiny sprouts which penetrate into the perivascular connective tissue; (b) Migration toward the stimulus of endothelial cells at the sprout tip; (c) Proliferation of the endothelial cells below the sprout; (d) Canalization, branching and formation of vascular loops, then of a functioning circulatory network; (e) Perivascular apposition of pericytes, and neosynthesis of basement membrane constituents by both the endothelial cells and the pericytes. Endothelial cell proliferation and migration coincide with limited laminin deposition, whereas cell differentiation and lumen formation coincide with further laminin deposition and type IV collagen deposition.

New microvessels grow by at least three mechanisms: (a) New sprouts bud from preexisting vessels; (b) Circulating endothelial progenitor cells participate in new vessel formation; (c) Endothelial cells in preexisting vessels bridge the lumen to form new vessels by intussusception. This latter postulated that the capillary network increases its complexity and vascular surface by insertion of a multitude of transcapillary pillars, a process called 'intussusception' (Djonov et al. 2000).

All three mechanisms depend upon loosening of preexisting endothelial cells from their junctions with each other which are maintained by proteins such as vascular endothelial (VE)-cadherin and platelet-endothelial cell adhesion molecule

D. Ribatti, *Angiogenesis and Anti-Angiogenesis in Hematological Malignancies*,
DOI 10.1007/978-94-017-8035-3_1, © Springer Science+Business Media Dordrecht 2014

Table 1.1 Main positive angiogenesis modulators

Vascular endothelial growth factor (VEGF)
Fibroblast growth factor-2 (FGF2)
Interluekin-8 (IL-8)
Platelet derived growth factor (PDGF)
Angiopoietin-1 (Ang-1)
Placental growth factor (PlGF)
Transforming growth factor beta (TGF-β)
Tumor necrosis factor alpha (TNF-α)
Hepatocyte growth factor (HGF)
Pleiotropin

Table 1.2 Main negative angiogenesis modulators

Angiostatin
Throbospondin1/2
Endostatin
Vasoinhibin
Vasostatin
Arresten
Canstatin
Tumstatin
Interferon α/β
Prolactin fragment
Fragment of platelet factor 4
Antithrombin fragment III

(PECAM); their junctions with contiguous pericytes, which are increased by angiopoietin-1 (Ang-1) and decreased by Ang-2; and their attachment to underlying basement membrane proteins which are governed by integrins such as $alpha_v beta_3$, and by a variety of local proteinases and their inhibitors. The endothelial cell loosening process may be aided by early dilation of microvessels which occurs prior to sprout formation and which is partly mediated by nitric oxide synthase (NOS).

Under physiological conditions, angiogenesis depends on the balance of positive and negative angiogenesis modulators within the vascular microenvironment (Tables 1.1 and 1.2) (Hanahan and Folkman 1996).

1.2 Tumor Angiogenesis

The first description of sprouting angiogenesis in tumor growth was reported by Ausprunk and Folkman in 1977, which indicated the following stages: (a) The basement membrane is locally degraded on the side of the dilated peritumoral postcapillary venule situated closed to the angiogenic stimulus; (b) Interendothelial contacts are weakened and endothelial cells migrate into the connective tissue; (c) A solid cord of endothelial cells form; (d) Lumen formation occurs proximal to the migrating front, contiguous tubular sprouts anastomose to form functionally capillary loops, parallel with the synthesis of the new basement membrane and the recruitment of pericytes.

Tumor angiogenesis goes through the same steps, but is uncontrolled and unlimited in time, and characterized by a 30/40-fold proliferative activity of endothelial cells (Ribatti and Vacca 2008). It is essential for tumor progression in the form of growth, invasion and metastasis because these develop through the transition from the avascular to the vascular phase.

The avascular phase has been studied by using tumor spheroids (e.g., of mouse melanoma B-16) in agar (Sutherland et al. 1971) and tumor implants (e.g., of Brown-Pearce carcinoma) into the anterior chamber of rabbit eye (Gimbrone et al. 1972); the human counterpart is *in situ* carcinoma and melanoma. In spheroidal tumors, the cell mass (volume) grows with the cube of the radius, whereas the surface area increases in proportion to its square. It follows that the growth of a tumor reaches a steady state when its area becomes too small to allow nutritional material to be supplied to its deep regions and the removal of metabolites. The spheroids are nourished solely by diffusible substances in the culture medium and thus reach a steady state when their mass is very small as many cells enter the system at the surface as are lost in depth by necrosis. Proliferative activity is very slow ("dormant" phase). The tumor has no metastatic potential and may remain in this phase indefinitely (Folkman and Grenspan 1975).

Conversely, if a tumor (e.g., the Brown-Pearce carcinoma) is implanted onto the rabbit iris, i.e., a site with an angiogenic potential, it is soon permeated by new vessels, its proliferative activity becomes exponential, all neoplastic cells continue to multiply and very few are lost from the system, its mass grows rapidly, and the steady state is reached at very high levels (4,000–16,000 times the original volume) (Gimbrone et al. 1972). It has also been observed that the mitotic index is inversely correlated with the distance between the tumor cell and the vessel, hence cells with the highest index are found within 1,000 μm from the capillary (Tannock 1968). Because O_2 diffuses to about 150 μm, it can be estimated that a cylinder with radius 150 μm centered around a 100 μm long capillary (made up of 20–100 endothelial cells) contains 10^4 viable tumor cells. It then follows that every time a capillary becomes one endothelial cell longer, it secures the existence and proliferation of 10^2 new tumor cells. It has been calculated that a tumor-induced capillary may elongate by about 800 μm/day, and thus support a growth of about 10^4 tumor cells/day (Folkman and Grenspan 1975). Angiogenesis is thus an amplification loop for tumor growth.

The vascular phase corresponds to locally invasive and metastatic tumors (Folkman et al. 1989), because the ingrowth of new capillaries provides the conditions for intravasation and colonization of other organs by tumor cells. In addition, tumor vessels are lined with a discontinuous endothelium that favours transmigration of these cells. To proceed, they must be capable of inducing a further angiogenic response. If not, they will behave like "dormant" micrometastases. The vascular phase also parallels tumor progression in terms of neoplastic changeover. In all these systems, the neoplastic or preneoplastic cell capable of inducing new vessels is ultimately responsible for the transition from the avascular to the vascular phase, i.e., for the escape from the "dormant" phase.

Tumor blood vessels display a markedly abnormal phenotype as well as altered genetic profile. All around and inside the tumor mass, they constitute a chaotic mixture of atypical, more dilated and tortuous blood channels, which are hierarchically disorganized as they lack the defining structural features of arterioles, capillaries, or venules (Ribatti et al. 2007). Tumor vessels form arteriovenous shunts, exhibit excessive branching, uneven diameters and, remarkably, increased permeability to macromolecules. The most consistent structural defects are related to endothelium. Tumor-associated endothelial cells proliferate 50–200 times faster than normal endothelial cells. The endothelium they form shows discontinuities or gaps that allow haemorrhage, and facilitate permeability of macromolecules and the traffic of tumor cells into the bloodstream. The basement membrane may have extra layers that exhibit no apparent association with endothelial cells or pericytes. Pericytes of tumor vessels are loosely associated with endothelial cells, have abnormal shape, paradoxically extend cytoplasmic processes away from the vessel wall, and have extra layers of loosely fitting basement membrane. The abnormal structure of tumor vessels is associated with altered gene and protein expression profiles in tumor endothelial cells (Aird 2009). Such findings provide evidence for the existence of a specific tumor endothelial cell transcriptional profile as distinct from the normal angiogenesis-related gene activation pattern.

Angiogenesis is the main process by means of which tumors create their own oxygen and nutrient supply and a route for systemic metastasis. In addition to angiogenesis, other mechanisms have been recognized to contribute to tumor vascularization. These include: (a) recruitment of circulating endothelial precursor cells (EPCs); (b) co-option of pre-existing blood vessels; and (c) vascular mimicry (Ribatti 2004). Bone marrow-derived EPCs have been proposed to be incorporated into tumor blood vessels although other studies indicate that their contribution to build-up tumor vessel endothelium may be modest. Cooption is illustrated in some forms of a highly malignant brain tumor, glioblastoma multiforme, in which tumor cells form cuffs that envelop normal brain blood vessels. Vascular mimicry is found in certain tumors, particularly ocular melanomas, and consists of spaces filled with red blood cells that are lined by tumor cells rather than endothelial cells. It is not clear whether the tumor cells truly "mimic" endothelial cells in their typical functions or whether they simply invade the vascular wall and become exposed to the blood flow as the result of endothelial cell apoptosis (Ribatti et al. 2007).

Solid tumors growth comprises an avascular and a subsequent vascular phase. If the second phase is dependent on angiogenesis and release of angiogenic factors, acquisition of angiogenic capability can be seen as an expression of the progression from neoplastic transformation to tumor growth and metastasis. The role of angiogenesis in the growth and survival of haematological malignancies has become evident since 1994, when Vacca et al. reported for the first time increased microvascular density (MVD) within the bone marrow of multiple myeloma (MM) versus monoclonal gammopathy of undetermined significance

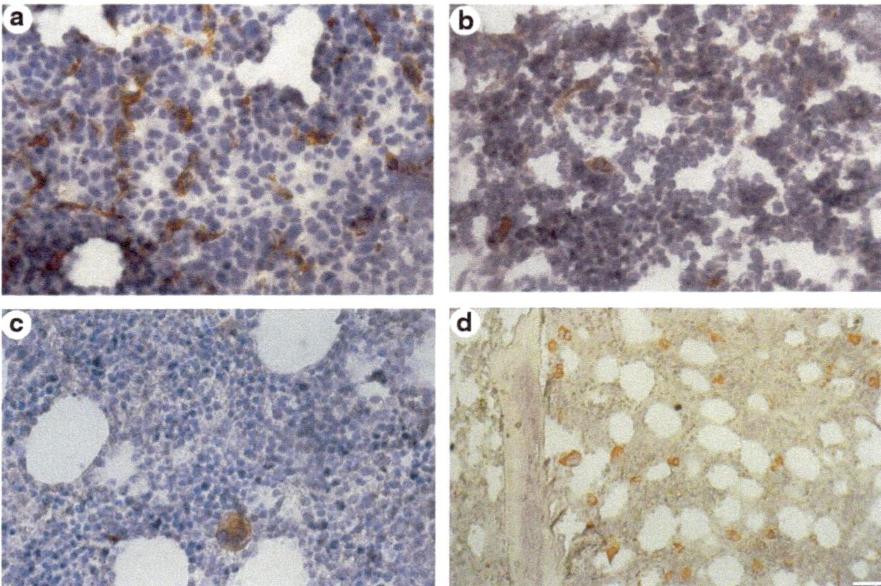

Fig. 1.1 Staining with factor VIII of bone marrow from patients with **a** MM at relapse, **b** MM at plateau, **c** MGUS, and **d** a control subject (patient with pernicious anemia). Note in **a** numerous microvessels, whereas in **b** a microvessel and some rare endothelial cell clusters and in **c** and **d** the lack of vessels in presence of strongly stained megakaryocytes. (Reproduced from Vacca et al. 1999a)

Table 1.3 Historical review of angiogenesis involvement in hematological malignancies

First evidence of bone marrow angiogenesis in multiple myeloma (Vacca et al. 1994)
First evidence of bone angiogenesis in B cell non Hodgkin lymphoma (Ribatti et al. 1996)
First evidence of bone marrow angiogenesis in acute lymphocytic leukemia (Perez-Atayade et al. 1997)
First evidence of angiogenesis involvement in the pathogenesis of B cell chronic lymphocytic leukemia (Molica et al. 1999)
High expression of VEGF in plasma cells, myeloid and monocyte precursors (Bellamy et al. 2001)
Detailed phenotypic, genetic and functional characterization of bone marrow endothelial cells from patients with multiple myeloma (Vacca et al. 2003b)

(MGUS) and in active (diagnosis, relapse, leukemic phase) versus non active (complete/objective response, plateau) MM (Fig. 1.1). Since 1994, several evidence have been accumulated confirming the role of angiogenesis in haematological tumors (Table 1.3).

Chapter 2
Angiogenesis in Multiple Myeloma

2.1 General Features of Multiple Myeloma

Multiple myeloma (MM) is the second most common hematological malignancy and is responsible for approximately 2 % of cancer death. Clonal expansion of malignant terminally differentiated, B-lymphocyte-derived plasma cells is characteristic of MM and results in excessive production of monoclonal immunoglobulins, thereby contributing to renal failure as well as other complications, such as hyperviscosity (Table 2.1). Diagnosis is characterized by the presence of monoclonal immunoglobulin, the presence of bone marrow plasmacytosis and typical skeletal lesions. Survival for patients with MM range from <6 months to more than 10 years based on disease stage and prognostic factors (Table 2.2). For several decades, the treatment of MM included combinations of chemotherapeutic agents with steroids for elderly patients and high-dose melphalan followed by autologous stem-cell transplantation after an induction with the combinations vincristine/doxorubicin/demethasone for patients <65–70 years.

In MM the neoplastic population consists of 3 clonally related compartments (a) a small stem-cell compartment, composed of totipotent cells that are self-maintaining and supply cells to the downstream compartments, (b) an expansion compartment or "growth fraction" (GF),whose proliferative activity expands the small population entering from the stem-cell compartment, (c) a differentiation compartment-the largest where proliferation stops and differentiation into mature plasma cells takes place (Billadeau et al. 1993). Cell loss by apoptosis and necrosis occurs in this compartment.

The kinetics of the stem-cell is unknown, but should be quite negligible, given the very small size of its compartment. Instead, we have plenty of information about the kinetics of the GF, the basic proliferative nucleus that contains cells in the G1, S, G2 and M phases, as well as resting GO cells, i.e., cells outside the proliferative cycle, but capable of joining it again following different stimuli (Drewinko et al. 1981). The balance between GF cell production and cell loss from the differentiation compartment determines growth, steady state or reduction of the myeloma cell mass.

D. Ribatti, *Angiogenesis and Anti-Angiogenesis in Hematological Malignancies*,
DOI 10.1007/978-94-017-8035-3_2, © Springer Science+Business Media Dordrecht 2014

Table 2.1 Multiple Myeloma staging system

Stage I	
1	Low M-component levels
2	Absent or solitary bone lesions
3	Normal hemoglobin, serum calcium, Ig levels (non-M component)
Stage III (any one or more of the following)	
1	High M-component
2	Advanced, multiple lytic bone lesions
3	Hemoglobin <8.5 g/dL, serum calcium >12 mg/dL
Stage II (overall values between I and III)	
Subclassification based on renal function: A, serum creatinine <2mg/dL; B, serum creatinine >2 mg/dL)	

Table 2.2 Diagnostic criteria for multiple myeloma

A	The diagnosis of myeloma requires a minimum of one major and one minor criteria or three minor criteria which must include (1) and (2). These criteria must manifest in a symptomatic patient with progressive disease
B	*Major criteria*
1	Plasmacytoma on biopsy
2	Marrow plasmocytosis (>30%)
3	M-component
C	*Minor criteria*
1	Marrow plasmacytosis (10–30%)
2	M-component: present but less than above
3	Lytic bone lesions
4	Reduced normal immunoglobulins

GF proliferative activity is measured by means of the labeling index (LI%), which identifies the S-phase (DNA synthesis) cell fraction (Durie and Salmon 1975). At diagnosis, the LI% is generally low (median = 1 % range 0–5) (Salmon 1975) and is calculated on the basis of the overall neoplastic population. If other methods are used and the LI% is solely related to the GF, which is generally small (mean <1 %, range <1–3.7) (Drewinko et al. 1981), it is much higher (median = 30 %, range 15–100), indicative of a very lively proliferation. This is confirmed by the fact that the myeloma cell mass doubling time is relatively long (5–15 months), with a huge cell loss (>90 %) given the remarkably large size of the differentiation compartment (Drewinko et al. 1981; Durie and Salmon 1975). It is then inferred that cell production by the GF is quite high. In short, the bone marrow myelomatous population consists of a small nucleus of rapidly proliferating cells, and a huge amount of non proliferating differentiated cells that are eventually lost.

MM is clinically diagnosed when the cellular mass is $>10^{12}$ cells -a stage reached following Gompertzian growth (Salmon and Seligman 1974). The initial growth is extremely fast (exponential), and then progressively reduced (LI% and GF decrease) according to a log factor, just as cell density continues to increase, until a steady state is reached (about 5–25 months after neoplastic changeover)

(Salmon and Seligman 1974). The length of this factor differs from one patient to another and determines whether the steady state is reached at high or low cell mass levels. When diagnosed, the plasma cell proliferative activity is extremely varied, depending on the steady state level (a) if the steady state has been reached, further growth is moderate (low LI% and GF), cell loss is high (>90 %), and the disease is poorly aggressive (non-active) (b) if the steady state is still a long way ahead, even though the cell mass is large enough (> 10^{12} cells) to elicit clinical symptoms, growth is exponential (high LI% and GF), cell loss is reduced (50–60 %), the tumor mass expands rapidly, and the disease is aggressive (active). There is no connection between LI% and the clinical stage (Durie et al. 1980).

In relapse which is usually detected when the cell mass is < 10^{12} cells, growth is exponential because the steady state is distant, and the disease is active. In the post-treatment stable remission, or plateau phase (> 50 % reduction of the M component for at least 6 months), the cell mass and the growing steady state have been reduced to a level allowing no further reduction, most cells are in G0 and growth is therefore limited (LI < 1 %, GF 10.5 %), and the disease is quiescent (non-active). While on treatment, after an initial and transient recruitment of G0 cells into the proliferative cycle and an increase in LI% (Salmon 1975) responsive patients exhibit growth conditions close to the plateau phase and the disease is non-active.

According to the Gompertzian growth pattern, the neoplastic population in MGUS reaches the steady state at very low cell mass levels (10^{10}–10^{11} cells). LI is very low (<1 %). The Gompertzian model relates active MM (exponential growth phase) to experimental solid tumors in the vascular phase, and non-active MM and MGUS (slow growth phase) to their avascular ("dormant") phase.

Mean survival is shorter in patients with LI 21 % (15 months) than in those with LI < 1 % (40 months), irrespective of the cell mass. In the Gompertzian pattern growth, MM with elevated LI% is still far from the steady state, since its mass is in the exponential portion of the growth curve. Patients rapidly respond to induction cytostatic treatment ("early responders"), but their response is short lived and followed by an equally rapid relapse due to massive recruitment of G0 plasma cells, and hence by shorter overall survival. Since LI% refers to the whole myelomatous population, not to GF alone, very small variations (e.g., by 1 %) really denote substantial changes in GF proliferative activity. This must be taken into account when forming a prognosis.

2.2 Angiogenesis in Multiple Myeloma

In 1994, Vacca and colleagues demonstrated for the first time that bone marrow MVD was significantly increased in MM compared to MGUS and moreover in active versus non-active forms.

The close association between angiogenesis and active MM indicates that it is the vascular phase of plasma cell tumors, and thus the counterpart to the vascular tumor implants of the rabbit iris and of locally invasive and metastatic solid tumors.

Conversely, MGUS and non-active MM represent the avascular phase, and correspond to tumor spheroids in agar, to tumor implants of the rabbit eye anterior chamber or to *in situ* tumors. Since most patients with active MM have a microvessel area 22 % and LI 21 %, a 2 % "vascular bone marrow threshold" seems to be required for this phase and rapid tumor growth.

The microvessel area and the LI% are closely associated with the MM activity phase and are mutually correlated. A disease in a given steady state (MGUS, non-active MM) can thus be supposed to at risk of progression towards the subsequent larger-mass steady state (active MM) if the bone marrow shows angiogenesis, and that this risk is greater the larger the microvessel area, namely 3.9 times higher for each 1 % increment. The risk is the same with 0.6 % increments of the LI. Microvessel area variations are therefore less restricted than those of LI% as a risk assessment parameter and could be a useful guide to prognosis just as LI% is utilized in plasma cell proliferative diseases. Angiogenesis correlates with LI%, but not with plasmacytosis, which mostly measures the compartment.

These findings as a whole suggest that, like solid tumor cells, highly proliferating-but not slowly proliferating nor differentiated myeloma cells, possess an angiogenic capability that could be expressed by the release of a variety of factors as follows: (a) Recruitment and activation of the bone marrow microenvironment to release interleukin-6, -8 (IL-6, IL-8), granulocyte-colony stimulating factor (G-CSF), granulocyte macrophage CSF (GM-CSF), tumor necrosis factor alpha (TNF-α), (b) Extracellular matrix proteolysis with fibroblast growth factor-2 (FGF-2) release, (c) Secretion of transforming growth factor beta (TGF-β) and IL-1β stimulating platelets to secrete platelet derived growth factor (PDGF). It should be noted that IL-6, IL-8, GM-CSF and IL-1β are mostly secreted in the MM active phase where angiogenesis is found. Because in turn angiogenesis favours plasma cell growth, it could thus complete a positive loop.

The progression of MGUS and non-active MM to active MM could take place when plasma cells become able to induce angiogenesis as described, resulting in formation of new vessels that themselves promote progression. Increased microvascular density and high bone marrow angiogenesis are an adverse prognostic factor in MM (Sezer et al. 2000; Iwasaki et al. 2002). CD34-positive microvessel areas in the bone marrow of patients with newly diagnosed MM correlate in a multivariate analysis with bone marrow plasma cell infiltration and b2-microglobulin (Sezer et al. 2001b). Moreover, after chemotherapy, MVD decreases significantly inpatients in complete or partial remission (Sezer et al. 2001b).

MM patients' bone marrow endothelial cells express and secrete higher amounts of the IL-8, interferon inducible T-cell-a chemoattractant (I-TAC), stromal cell derived factor 1 alpha (SDF-1α), and monocyte chemotactic protein-1 (MCP-1) than healthy umbilical vein endothelial cells (HUVECs) (Pellegrino et al. 2005). The direct production of osteopontin, the expression and activity of its major regulating gene Runx2/Cbfa1 in MM cells, and the potential role of osteopontin in plasma cell induced bone marrow angiogenesis have been elucidated (Colla et al. 2005; Cheriyath et al. 2005).

Human MM cell lines express and secrete Ang-1, but not its antagonist Ang-2 (Giuliani et al. 2003). Ang-1 alone is expressed in about 47% of newly diagnosed MM patients in stages I–III. Moreover, microarray comparison of the gene expression profiles of bone marrow plasma cells and extramedullary plasmacytoma cells has shown an eightfold upregulation of Ang-1 expression in the latter (Hedvat et al. 2003). Evaluation of the density and number of vessels per field in bone marrow biopsies has revealed a significant increase of angiogenesis in Ang-1 positive as opposed to Ang-1-negative patients (Giuliani et al. 2003). A 5.8-fold increase of Ang-1 transcript was detected in MM associated with higher angiogenesis (Munshi et al. 2004). The presence of anti-Tie2 (Ang-1 receptor) blocking antibody completely abolished vessel formation induced by conditioned medium of several human MM cell lines (Giuliani et al. 2003). Multiple myeloma plasma cells influence the expression of Angs and Tie2 receptor in the bone marrow milieu. Several human MM cell lines upregulate the expression of Tie2 by endothelial cells at both the mRNA and the protein level (Giuliani et al. 2003). This *in vitro* effect has been confirmed *in vivo* by the finding of overexpression of Tie2 in isolated bone marrow MM endothelial cells as compared to HUVECs (Vacca et al. 2003a). Nakayama et al. (2004) showed that mast cells promote the growth of plasma cell tumors through secretion of Ang-1, which stimulates angiogenesis in conjunction with tumor-derived vascular endothelial growth factor (VEGF).

The angiogenic switch was owing to an increase in FGF-2 production and secretion by plasma cells, probably caused by expression of oncogenes (c-myc, c-fos, c-jun, ets-1) coding for angiogenic factors, and activated as a consequence of immunoglobulin translocations and genetic instability of plasma cells (Vacca et al. 2001a).

Accordingly, by using the 5T22 murine model for MM, Asosingh et al. (2004) showed that the switch is preceded by the expression of mRNA for VEGF and secretion of protein by plasma cells. Furthermore, they demonstrated that the angiogenic switch was preceded by an increase in the percentage of CD45 negative MM cells with high levels of VEGF secretion.

Other studies, however, showed that the expression levels of VEGF, FGF-2, and their receptors were similar among plasma cells from MGUS, smoldering MM and newly-diagnosed MM, suggesting that increasing angiogenesis from MGUS to MM is, at least in part, explained by increased tumor burden rather than increased expression of VEGF/FGF-2 by single plasma cells (Kumar et al. 2004). Another possibility is that the inhibition of angiogenesis in MGUS is lost with progression, hence that the switch from MGUS to MM may involve a loss of an anti-angiogenic activity (Kumar et al. 2004). We have demonstrated that in the chick embryo chorioallantoic membrane (CAM) assay MM endothelial cells induced an angiogenic response comparable to that of FGF-2, while reverse transcriptase-polymerase chain reaction (RT-PCR) demonstrated that the expression of endostatin mRNA detected in MM treated CAM was significantly lower respect to control CAM (Mangieri et al. 2008). These data suggest that angiogenic switch in MM may involve loss of an endogenous angiogenesis inhibitor, such as endostatin.

Hose et al. (2009) substained that "angiogenesis may not be critical for MM pathogenesis, but just an epiphenomenon driven by the accumulation of malignant plasma cells and a production of pro-angiogenic cytokines that a have dual role as growth and survival factors for MM cells".

2.3 Angiogenic Cytokines and Multiple Myeloma Progression

Interleukin-6 is a major growth, survival factor and anti-apoptotic factor for MM plasma cells. It is mainly produced by bone marrow stromal cells and also by plasma cells themselves. It is angiogenic (Motro et al. 1990). Elevated serum IL-6 receptor concentrations are associated with a poor prognosis in MM (Pulkki et al. 1996). Contact of plasma cells with bone marrow stromal cells upregulates the latter's transcription and secretion of IL-6 which are further enhanced by cytokines (particularly TGF-β) secreted by plasma cells (Urashima et al. 1996). Interleukin-6 stimulates VEGF secretion via an IL-6 receptor expressed by plasma cells (Cohen et al. 1996; Bellamy et al. 1999; Dankbar et al. 2000; Gupta et al. 2001). Similarly, stimulation of endothelial cells and bone marrow stromal cells with VEGF induced a significant and dose-dependent increase in IL-6 secretion.

The secreted isoforms of VEGF, such as VEGF121, VEGF145 and VEGF165, as well as the ECM and surface-bound VEGF189 and VEGF206 are produced by both MM cell lines and a patient's bone marrow plasma cells (Figs. 2.1 and 2.2) (Bellamy et al. 1999; Podar et al. 2001). VEGF acts as an autocrine inducer of growth and chemotaxis via VEGF receptor-1 (VEGFR-1) (Podar et al. 2001). It increases IL-6 production by bone marrow stromal cells via VEGFR-2 and thus forming a paracrine loop for tumor growth (Dankbar et al. 2000) and angiogenesis. Moreover, adhesion of plasma cells to bone marrow stromal cells increases VEGF secretion by both cell types (Hideshima et al. 2005), and so enhances angiogenesis. Vascular endothelial growth factor production by plasma cells is also regulated by TNF-α of bone marrow stromal cells (Neufeld et al. 1999). VEGF and its receptors VEGFR-2 and VEGFR-3 are highly expressed by bone marrow MM endothelial cells and plasma cells at both mRNA and protein levels compared to MGUS endothelial cells (Figs. 2.1, 2.3 and 2.4) (Vacca et al. 2003a; Ria et al. 2004). Elevated serum VEGF in MM patients is correlated with both increased angiogenesis in MM bone marrow and higher plasma cell LI (Vacca et al. 1999a; Rajkumar et al. 2000b).

Vincent et al. (2005) have shown that VEGF released by fetal bone marrow stromal cells interacts with VEGFR-1 of MM plasma cells in a paracrine fashion. Moreover, a neutralizing anti-VEGFR-1 monoclonal antibody, but only partially, blocks plasma cell motility *in vitro*, indicating that the functional VEGF-A/VEGFR-1 interaction is essential for MM homing and migration, and that bone marrow stromal cells secrete other cytokines and/or chemokines that promote plasma cell migration irrespective of the VEGF-A/VEGFR-1 signaling. Vascular endothelial growth factor may also mimic the role of M-CSF by increasing osteoclast development (Niida et al. 1999) and stimulating resorptive activity (Nakagawa et al. 2000), so that bone destruction is mediated via more than one route. VEGF triggers a signaling cascade

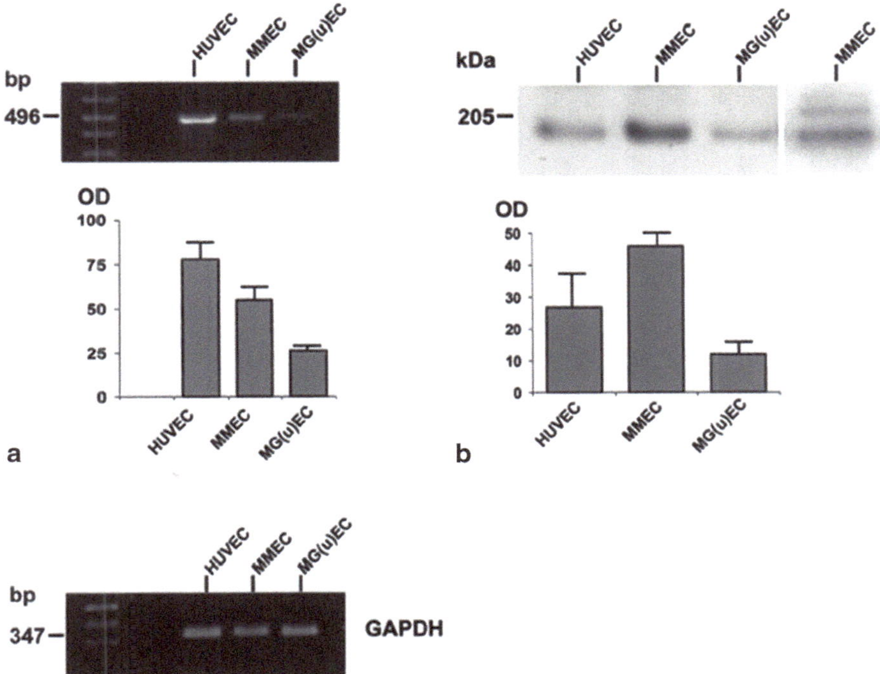

Fig. 2.1 Parallel expression of VEGF-A at mRNA (**a**), and protein (**b**), levels. Measurement by ELISA of the VEGF-A released into endothelial cell conditioned media. Histograms show mean ±1 SD in the groups of patients and HUVEC. (Reproduced from Ria et al. 2004)

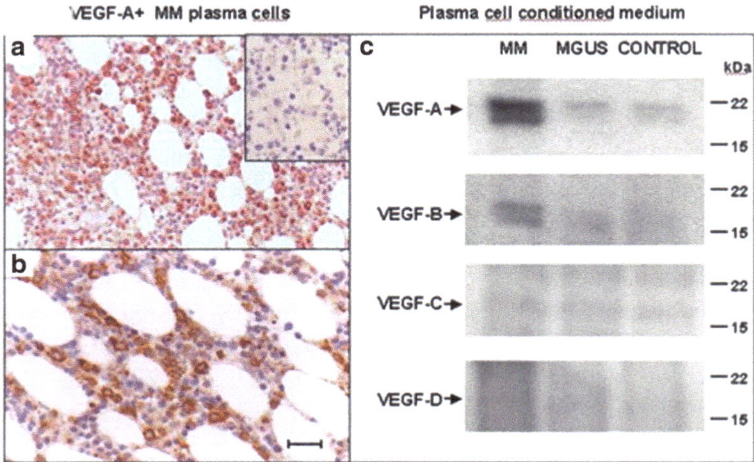

Fig. 2.2 Staining with VEGF-A of bone marrow from a patient with MM (**a**, **b**) insert: the negative control obtained with rabbit pre-immune serum replacing the VEGF-A antibody (**c**). Western blot analysis for VEGF homologs of plasma cell culture medium from a representative MM patient, MGUS patient and control subject. (Reproduced from Vacca et al. 2003a)

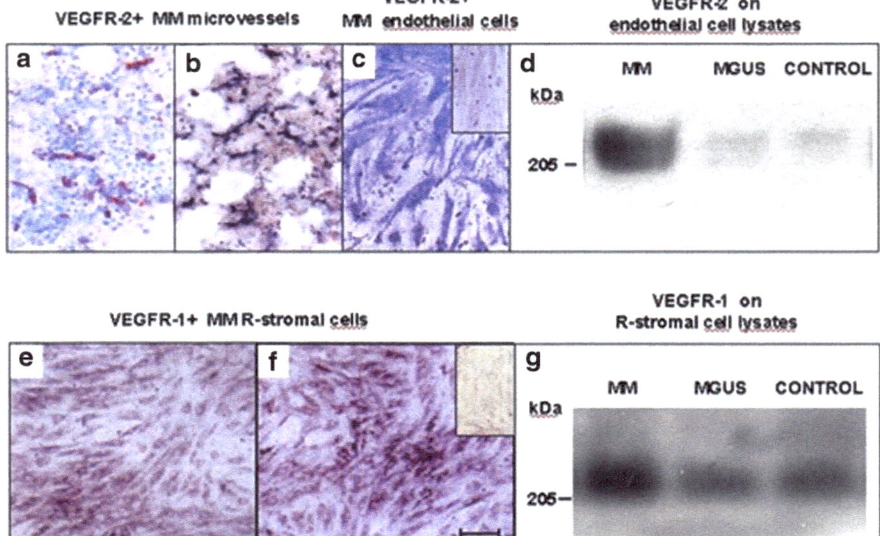

Fig. 2.3 Staining with VEGFR-2 of a bone marrow from a patient with MM showing numerous thin microvessels (**a**). *In situ* hybridization for VEGFR-2 mRNA highlighting neovessels (**b**), and enriched endothelial cells (**c**), of the MM patient. Insert: RNAse-treated cells as the negative control. Western blot analysis of the VEGFR-2 of endothelial cell lysates from the patients with MM, MGUS, and the control subject, showing enhanced expression in MM (**d**). Immunocytochemical staining with VEGFR-1 (**e**) and *in situ* hybridization fro VEGFR-1 mRNA of residual-stromal cells from the patient with MM (**f**). Insert: RNAse-treated cells as the negative control. Western blot analysis for the VEGFR-1 of residual-stromal cell lysate from the patient with MM, and the patient with MGUS and the control subject (**g**). The expression of VEGFR-1 is higher in MM. (Reproduced from Vacca et al. 2003a)

in plasma cells, including the extracellular-signal regulated kinases (ERK) pathway, which mediates cell growth, and the phosphatidylinositol-3 kinase/protein kinase C (PI3 K/PKC) -dependent cascade that mediates migration (Podar et al. 2001). Le Gouill et al. (2004) have demonstrated that VEGF upregulates expression of anti-apoptotic proteins, including myeloid cell leukemia, survivin, and cellular inhibitor of apoptosis protein. Furthermore, free bovine serum (starvation)-induced apoptosis in plasma cells is partially prevented by VEGF, confirming that this is a potent anti-apoptotic cytokine in MM (Lichtenstein et al. 1995; Chauhan et al. 1997a, b; Tu et al. 2000). Bone marrow stromal cells secrete also VEGF-C and VEGF-D (Vacca et al. 2003a). VEGF is thus prevalently produced by plasma cells and stimulates the proliferation and chemotaxis of MM endothelial cells via VEGFR-2 and of bone marrow stromal cells via VEGFR-1. Activation of the latter results in VEGF-C and VEGF-D production that stimulates plasma cell growth via VEGFR-3. Increased proliferation and chemotaxis displayed by endothelial cells and bone marrow stromal cells in response to plasma cell conditioned medium is not fully abolished by addition of anti-VEGF antibody, suggesting the secretion of other angiogenic cytokines (Vacca et al. 2003a).

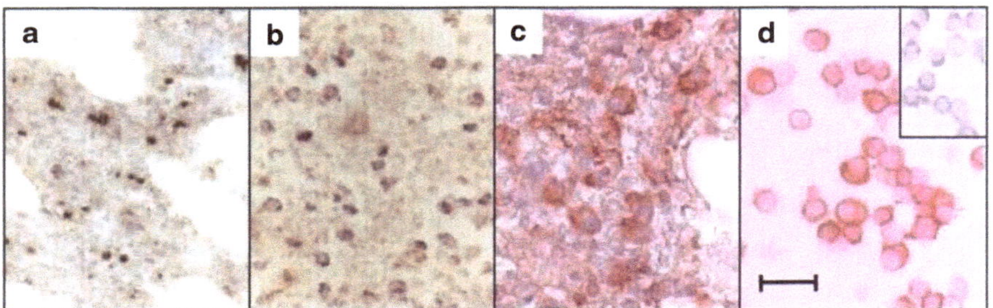

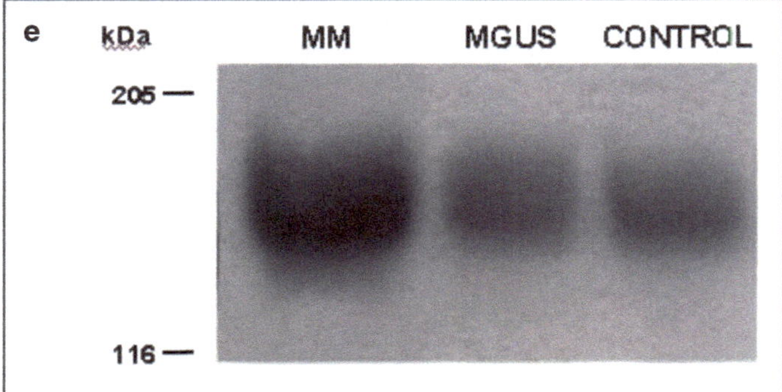

Fig. 2.4 *In situ* hybridization for VEGFR-3 mRNA of bone marrow from two patients with MM (**a**, **b**) and staining with VEGFR-3 of bone marrow (**c**) and enriched plasma cells (**d**), of the patient. Inset: the negative control obtained with rabbit preimmune serum replacing the VEGFR-3 antibody. Western blot analysis for the VEGFR-3 of plasma cell lysates from patient with MM, a patient with MGUS, and a control subject (**e**). The VEGFR-3 expression is higher in MM. (Reproduced from Vacca et al. 2003a)

Plasma cells are source of FGF-2 in the bone marrow of patients with active MM (Vacca et al. 1999; Bisping et al. 2003), and FGF-2 concentrations in marrow aspirates and peripheral blood correlate with MM activity (Di Raimondo et al. 2000; Sezer et al. 2001; Sato et al. 2002). MM plasma cells induced an intense angiogenic response *in vivo* the chick embryo CAM assay (Fig. 2.5) Ribatti et al. 2003a). An anti-FGF-2 antibody significantly inhibited angiogenesis induced *in vivo* in the CAM assay by MM plasma cell culture medium (Fig. 2.6), (Vacca et al. 1999a). Bisping et al. (2003) have investigated the expression of high affinity FGF receptor (FGFR-1) through R-4 in RPMI-8226 and U266 MM cell lines and patients' plasma cells, as well as in bone marrow stromal cells. Moreover, they demonstrated that FGF-2 induces a time- and dose-dependent increase in IL-6 secretion by bone marrow stromal cells, suggesting that FGF-2 is both an angiogenic growth factor in MM via IL-6 and a supporter of plasma

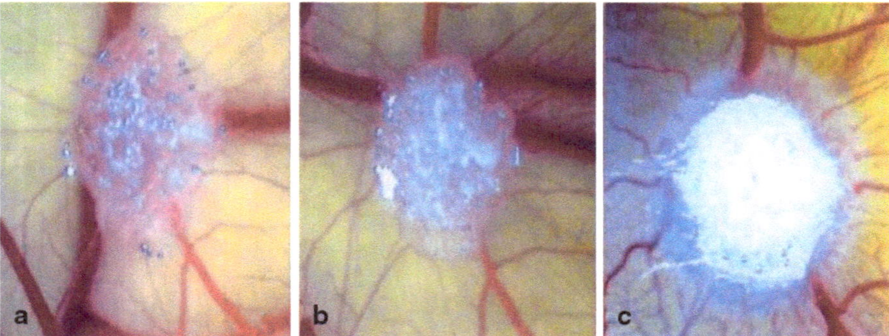

Fig. 2.5 Time-course of the macroscopic appearance of a CAM implanted at day 8 (**a**), with a sponge loaded with 18,000 plasma cells of an active MM patient. Note that, whereas on day 9 (**b**), no vascular reaction is detectable, on day 12 (**c**), numerous allantoic vessels develop radially towards the implant in a "spoked-wheel" pattern. (Reproduced from Ribatti et al. 2003a)

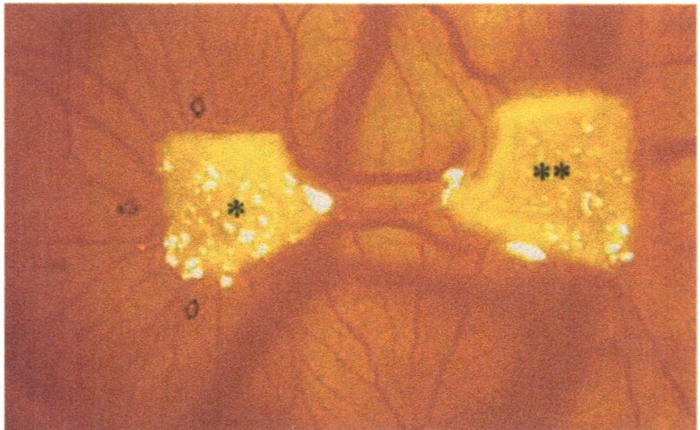

Fig. 2.6 Macroscopic picture of a CAM at day 12 of incubation implanted simultaneously at day 8 with a sponge loaded with the plasma cell culture medium of an active MM patient alone (*) and with a second sponge loaded with the same culture medium added with an anti-FGF-2 antibody (**). Note the angiogenesis toward the one-asterisk sponge (some neovessels are *arrow-headed*), and its inhibition by the anti-FGF-2 antibody. (Reproduced from Vacca et al. 1999a)

cell growth and survival by paracrine stimulation of the IL-6 released by bone marrow stromal cells.

TNF-α mediates upregulation of adhesion molecules of plasma cells lymphocyte function associated antigen-1 and very late antigen-4 (LFA-1 and VLA-4) and bone marrow stromal cells intercellular cell adhesion molecule-1 and vascular cell adhesion molecule-1 (ICAM-1 and VCAM-1), and thus enhances heterotypic adhesions and activates IL-6 secretion by bone marrow stromal cells (Hideshima et al. 2001a). Moreover, TNF-α secreted by plasma cells induces NF-kappa B (NF-kB) -dependent upregulation of adhesion molecules on both MM and bone marrow stromal cells (Hideshima et al. 2001b), thereby increasing the binding of MM to bone marrow stromal cells with associated cell adhesion mediated-drug resistance and induction of IL-6, insulin-like growth factor (IGF-1) and VEGF secretion by bone marrow stromal cells (Chauhan et al. 1996; Hideshima et al. 2001a, 2002b). Activation of plasma cells via CD40 ligand, a TNF-α family member, induces VEGF secretion in bone marrow stromal cells, which mediates MM cell homing and migration, as well as angiogenesis (Tai et al. 2002).

Hepatocyte growth factor/scatter factor (HGF/SF) is yet another angiogenic factor identified in the culture medium of both human MM cell lines (Børset et al. 1996b) and freshly isolated plasma cells (Seidel et al. 1998). Serum levels of this factor were high in 43 % of patients at diagnosis, fell to normal levels if a response to induction therapy was achieved and rose again upon relapse (Di Raimondo et al. 2000). HGF/SF stimulates the proliferation of plasma cells and protect them from apoptosis. C-Met, the receptor of HGF/SF, is expressed by the majority of MM cells (Derksen et al. 2003). A positive correlation exists between HGF/SF, syndecan-1 expression and bone marrow angiogenesis in MM (Andersen et al. 2005). Homeobox B7 (HOX) overexpression by MM cells increased their pro-angiogenic properties (Fig. 2.7; Storti et al. 2011).

Still other factors may be responsible for bone marrow angiogenesis in MM: TGF-β and IL-1β;IL-8 (Shapiro et al. 2001); G-CSF, GM-CSF (Hallek et al. 1998) secreted by bone marrow stromal cells recruited and activated by plasma cells. IL-8, GM-CSF and IL-1β, indeed, are mostly secreted in the MM active phase (Hallek et al. 1998).

MicroRNA (miRNAs) regulate many cellular functions including cell proliferation, apoptosis, and differentiation. It has been established that miRNA expression profile differ between normal tissues and the derived tumors and between the different tumor types. Roccaro et al. (2009) demonstrated that miRNA-15a and -16 are critical regulators of MM pathogenesis both directly by targeting clonal plasma cells, and indirectly by reducing bone marrow neoangiogenesis and the interaction between tumor cells and bone marrow milieu (Fig. 2.8).

2.4 Signaling Pathways

Multiple myeloma plasma cell binding to bone marrow stromal cells upregulates cytokines secretion from both cell types. These interactions subsequently activate a pleiotropic cascade of proliferative/antiapoptotic signaling pathways: PI3 K/AKT; I kappa B kinase (IKK)/(NF-kB); Ras/Raf/mitogen-activated protein kinase (MAPK); mitogen-induced extracellular kinase (MEK)/ERK; Janus kinase (JAK)2/signal transducers and activators of transcription (STAT)3. Downstream sequelae include: cytoplasmic sequestration of proapoptotic forkhead transcription factor (FKHR); upregulation of cyclin-D and antiapoptotic Bcl-2 family members and increased activity of telomerase. Interleukin-6 activates the JAK2/STAT3 pathway in plasma cells, resulting in the overexpression of the antiapoptotic proteins Bcl-xL and Mcl-1, and inhibition of CD95 (Fas)-induced apoptosis (Catlett-Falcone et al. 1999; Jourdan et al. 2003).

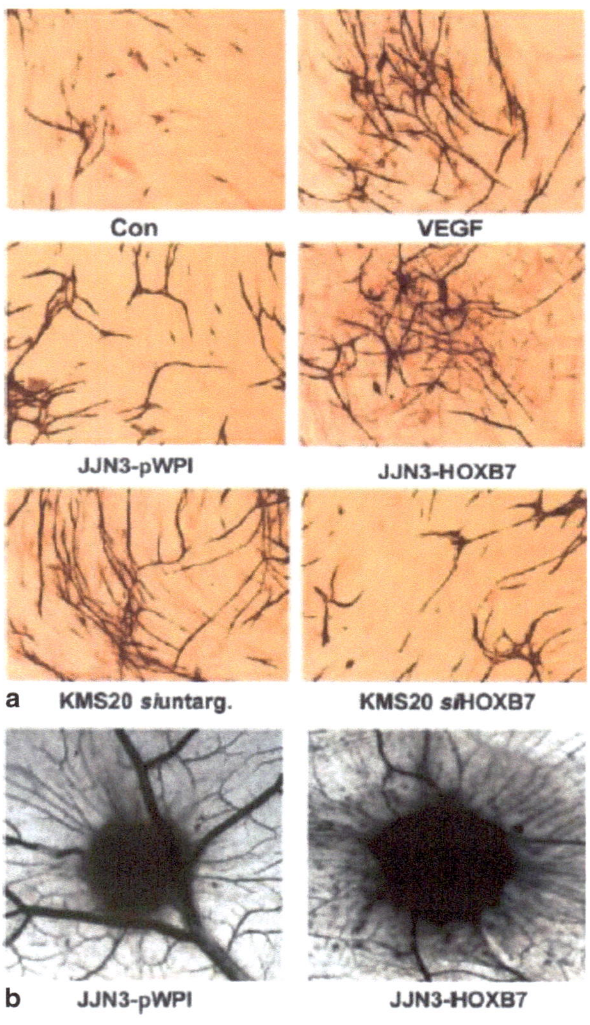

Fig. 2.7 HOXB7 overexpression by MM cells increases their pro-angiogenic properties (**a**). Endothelial-like cells were incubated with DMEM medium at 10% FBS (Con) or with DMEM medium at 10% PBS plus VEGF of with culture medium of human myeloma cell line JJN3-pWPI of with JJN3 stably transfected with a lentivirus vector carrying HOBX7 (JJN3-HOXB7)/D-MEM at 10% FBS or human myeloma cell line KMS-20 *si*untarg or mixed with *si*RNA against HOXB7 (KMS20 *si*HOXB7) (**b**). Chick embryo CAMs were treated with gelatin sponges loaded with culture medium of JJN3-pWPI or of JJN3-HOXB7. (Reproduced from Storti et al. 2011)

Plasma cells become independent of the IL-6/JAK2/STAT3 pathway when co-cultured with bone marrow stromal cells. This emphasizes that the crosstalk between plasma cells and bone marrow stromal cells stimulates additional IL-6-independent pathways. In fact, also IL-6 triggers plasma cells proliferation also via the Ras/Raf/MEK/MAPK cascade and protects against dexamethasone-induced apoptosis by Pl3 K/AKT signaling (Ogata et al. 1997; Tu et al. 2000; Hideshima et al. 2001a; Hu et al. 2003). Mcl-1 is tightly regulated by IL-6. It is essential for the survival of plasma cells *in vitro* (Zhang et al. 2002), and overexpressed by those from patients with relapse or poor prognosis (Wuillèlme-Toumi et al. 2005).

Insulin-like growth factor-1 induces proliferation, survival and drug resistance of plasma cells via Ras/MAPK and Pl3 K/AKT signaling cascades; it is a more potent

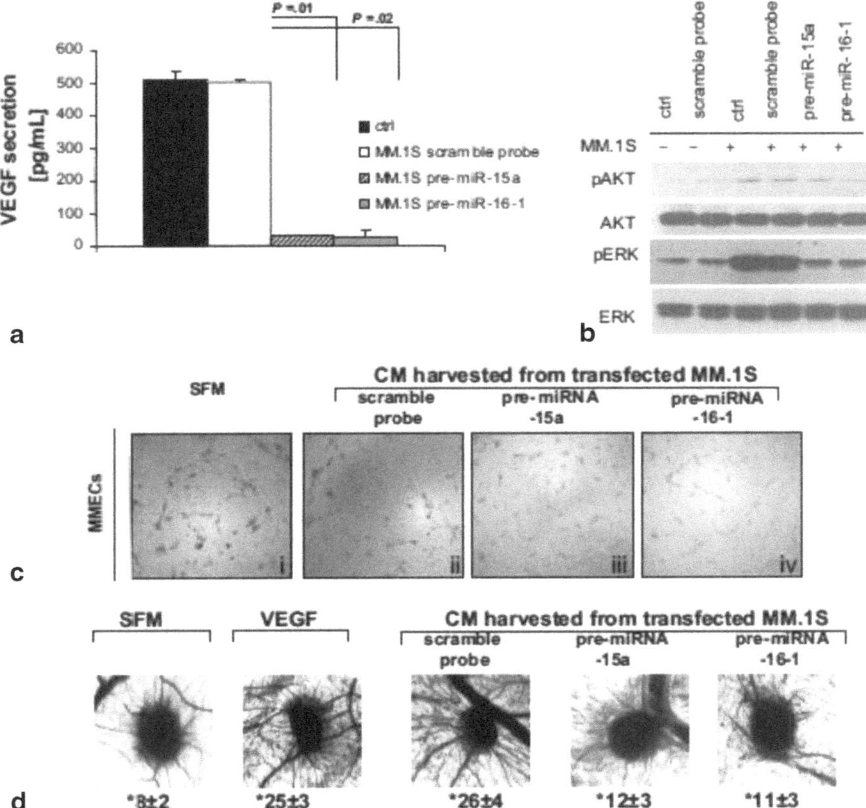

Fig. 2.8 miRNA-15a and -16 target endothelial cells *in vitro* and *in vivo* (**a**) VEGF concentrations were measured in triplicate, by ELISA, in conditioned media obtained from multiple myeloma cells (pre-miRNA-15a–, -16-1 probe–,control probe–transfected MM.1S) (**b**). HUVECs were cultured in presence or absence of multiple myeloma cells (pre-miRNA-15a–, -16-1 probe–, control probe–transfected MM.1S). Non transfected MM.1S cells were used as control (ctrl). Whole cell lysates were subjected to Western blotting using anti–phospho(p)-AKT, -AKT, p-ERK, -ERK, and-actin antibodies (**c**). Multiple myeloma endothelial cells were seeded on top of extracellular matrix, in presence of culture medium obtained from either control probe–, pre-miRNA-15a, or -16-1–transfectedMM.1S cells. Serum-free medium (SFM) was used as control. Tube formation was assessed using an inverted light microscope. Photographs are representative of 3 independent experiments (**d**). Angiogenic responses induced by gelatin sponges loaded with culture medium obtained from either control probe–, pre-miRNA-15a, or -16-1–transfected MM.1S cells, in the *in vivo* CAM model. SFM and VEGF (10 ng/mL) were used as negative and positive control, respectively. Vessel counts at the sponge-CAM boundary are indicated below each image. (Reproduced from Roccaro et al. 2009)

inducer of AKT signaling and NF-kB activation than IL-6, and confers protection against dexamethasone mediated apoptosis (Tu et al. 2000; Qiang et al. 2002). Insulin-like growth factor-1 also attenuates the anti-MM activity of other antitumor drugs, including cytotoxic chemotherapy and proteasome inhibitors (Mitsiades et al. 2002, 2004).

Vascular endothelial growth factor induces modest proliferation and pronounced migration via MEK/MAPK and P13 K/PKC signaling pathways, respectively (Podar et al. 2001), but not with ERK activation(Podar et al. 2002). It is abrogated by neutralizing specific antibody, but not by MAPK inhibitors. Le Gouill et al. (2004) showed that VEGF upregulates Mcl-1 expression in MMcell lines and MMplasma cells; conversely, a pan-VEGF inhibitor inhibits VEGFinduced upregulation of Mcl-1 associated with decreased proliferation and induction of apoptosis (Le Gouill et al. 2004). Stromal cell-derived factor-1 promotes proliferation, induces migration and protects against dexamethasone induced apoptosis via activation of the MAPK, P13 K/AKT, and NF-kB signaling cascades, respectively (Hideshima et al. 2002b).

2.5 Invasive Ability of Plasma Cells

Enriched bone marrow plasma cells expressed matrix metallprotinase-2 (MMP-2), whereas the expression of MMP-9 was less intense and inconstant, and secreted both enzymes into the culture medium (Vacca et al. 2000). Barillé et al. (1997) have shown production of MMP-9 by several MM cell lines and freshly isolated bone marrow plasma cells and demonstrated that bone marrow stromal cells constitutively produce MMP-1 and MMP-2. Furthermore, IL-1β and TNF-α, produced by MM plasma cells and bone marrow stromal cells, respectively (Di Raimondo et al. 2000) enhance MMP-1 production. Given the ability of MMP-1, MMP-2 and MMP-9 to degrade collagens and fibronectin, the data suggest that plasma cells of active MM patients are especially capable of invading both the stroma and the subendothelial basement membrane. In addition, extracellular matrix degradation brings about the release of its stored angiogenic factors.

Fresh bone marrow plasma cells from MM patients are able to interact with immobilized vitronectin and fibronectin through αvβ3 integrin (Vacca et al. 2001b). This interaction allows their prompt adhesion to the substratum and rapidly induces their recruitment of the β3 integrin subunit together with the cytoskeletal components vinculin and paxillin and results in an increase in MMP-2, MMP-9 and urokinase plasminogen activator (uPA) secretion. VEGF stimulates migration of an MM cell line and cell attachment to fibronectin via α1 integrin and PKC activation (Podar et al. 2002). These data imply that plasma cell adhesion to the fibronectin and vitronectin and subsequent stimulation of cell proliferation, migration and protease secretion are enhanced by several integrin complexes.

2.6 Multiple Myeloma Endothelial Cells

Isolated endothelial cells from the bone marrow of patients with active MM displayed antigenic, functional, genetic, and morphological features indicative of ongoing neovascularization (Vacca et al. 2003b). They secrete FGF-2, VEGF, MMP-2, and MMP-9 and are a heterogeneous population with a dissimilar expression of antigens, as evaluated by fluorescent activating cell sorter (FACS).

Long-term cultured endothelial cells from the bone marrow of patients with MM were compared with endothelial cells from patients with MGUS and HUVECs as their normal quiescent counterpart (Vacca et al. 2003b). MM endothelial cells showed enhanced morphological activation compared to HUVECs and expressed 5–70 times higher levels of several vascular markers (Tie2, VEGF, VEGFR-2, FGFR-2, CD105-endoglin, and VE-cadherin) than HUVECs (Ria et al. 2004). They produced 4–40 times more FGF-2 and VEGF, and 3–5 times more MMP-2 and MMP-9 than HUVECs. They gave a growth advantage over HUVECs because they quickly formed a closely knit capillary plexus with multicentric junctions and sprouts, whereas the plexus of the other endothelial cells was loose with few junctions and sprouts (Fig. 2.9) (Ria et al. 2004). Multiple myeloma endothelial cells induced an intense angiogenic response in the CAM. On a cDNA microarray, MM endothelial cells displayed upregulation of angiogenic genes (VEGF and FGF isoforms, HGF/SF, Tie2, TGF-β, Gro-α chemokine, fibronectin-1, hypoxia inducible factor 1 alpha (HIF-1α), ETS-1, ID3, and osteopontin) compared to HUVECs.

MM endothelial cells alone displayed a VEGF-dependent autocrine growth loop (Ria et al. 2004) owing to high VEGF and VEGFR-2 expression, constitutive autophosphorylation in both VEGFR-2 and the associated kinase ERK-2, and inhibition of proliferation, capillarogenesis and phosphorylation by neutralizing anti-VEGF and anti-VEGFR-2 antibodies. Overall findings seem to point to a tumoral phenotype of MM endothelial cells: either plasma cells and MM endothelial cells derive from a common tumor precursor cell, or MM endothelial cells are hybrids of plasma cells with existing endothelial cells. These hypotheses have been demonstrated in endothelial cells of B-cell lymphomas (Streubel et al. 2004).

Recently, Basile et al. (2013) demonstrated that pentraxin 3 affects both MM endothelial cell and fibroblast functional activities (Fig. 2.10). Moreover, pentraxin 3 is able to affect the angiogenic capability of both MM endothelial cells and plasma cells (Fig. 2.11).

A comparative gene expression profiling of MM endothelial cells and MGUS endothelial cells with Affymetrix U133 A arrays has been provided (Fig. 2.12; Ria et al. 2009). Twenty-two genes were found differentially expressed (14 down-regulated and 8 up-regulated) at relatively high stringency in MM endothelial cells compared with MGUS endothelial cells. Deregulated genes are mostly involved in ECM formation and bone remodeling, cell adhesion, chemotaxis, angiogenesis, resistance to apoptosis, and cell-cycle regulation. Validation was focused on *DIRAS3*, *SERPINF1*, *SRPX*, *BNIP3*, *IER3*, and *SEPW1* genes, which were not previously found to be functionally correlated to the overangiogenic phenotype of MM endo-

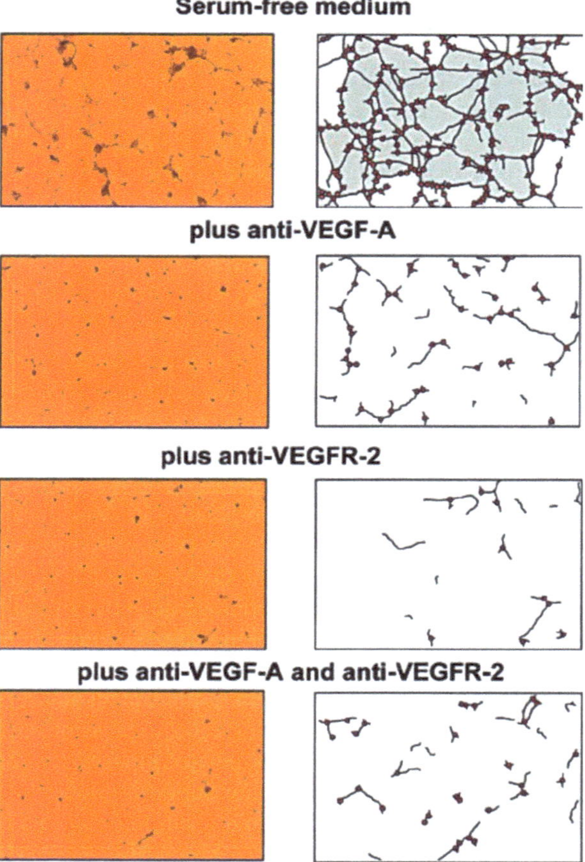

Fig. 2.9 Multiple myeloma endothelial cells were seeded on Matrigel in serum free medium. After 12 h of incubation, their three dimensional organization was examined planimetrically by a computed image analysis. At left, multiple myeloma endothelial cells formed branching anastomosing tubes with multicentring junctions resulting in a closely-knit capillary mesh, whose number of areas (*shaded areas*), vessel length and number of mesh branching points shown at right were 44, 5860 mm and 217. When anti-VEGF-A or anti-VEGFR-2 MoAb were added, the areas disappeared, length was reduced to 472 mm and 201 mm and points to 40 and 9 respectively. No additive effect was observed. (Reproduced from Ria et al. 2004)

thelial cells. Small interfering RNA for three up-regulated genes (*BNIP3*, *IER3*, and *SEPW1*) affected critical MM endothelial cell functions mediating the cell overangiogenic phenotype, that is proliferation, apoptosis, adhesion, and capillary tube formation.

Four proteins were found overexpressed in MM endothelial cells: filamin A, vimentin, α-crystallinB and 14–3-3ζ/δ protein (Fig. 2.13; Berardi et al. 2012). Their expression was enhanced by VEGF, FGF-2, HGF, and MM plasma cell conditioned medium and their silencing RNA knockdown affected MM endothelial cells an-

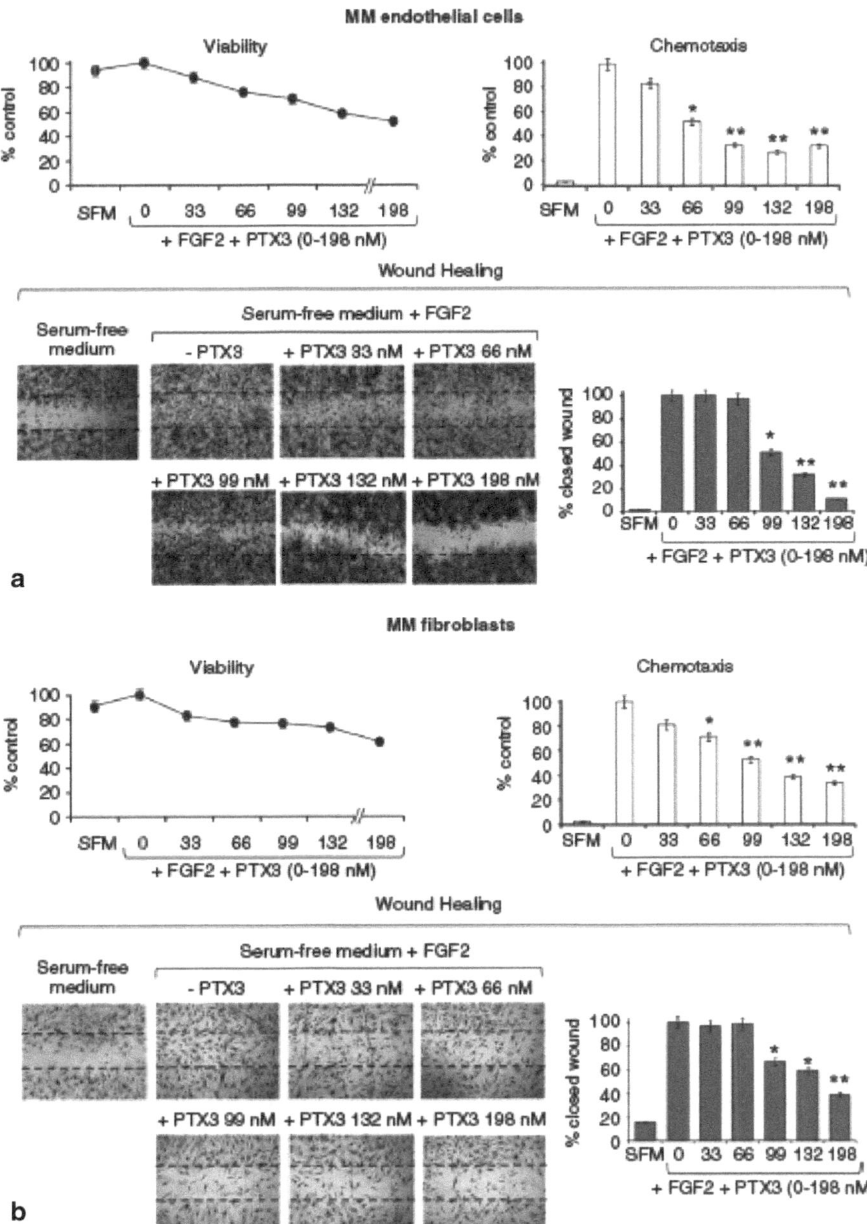

Fig. 2.10 PTX3 affects multiple myeloma (MM) endothelial cell (EC) and fibroblast (FB) functional activities. Viability, chemotaxis, and wound healing assays were performed on ECs (**a**), and FBs (**b**), isolated from the bone marrow of 27 MM patients and cultured in triplicate in serum-free medium (SFM) and SFM supplemented with FGF2 and PTX3 doses. Wound healing assays from a representative patient are shown. (Reproduced from Basile et al. 2013)

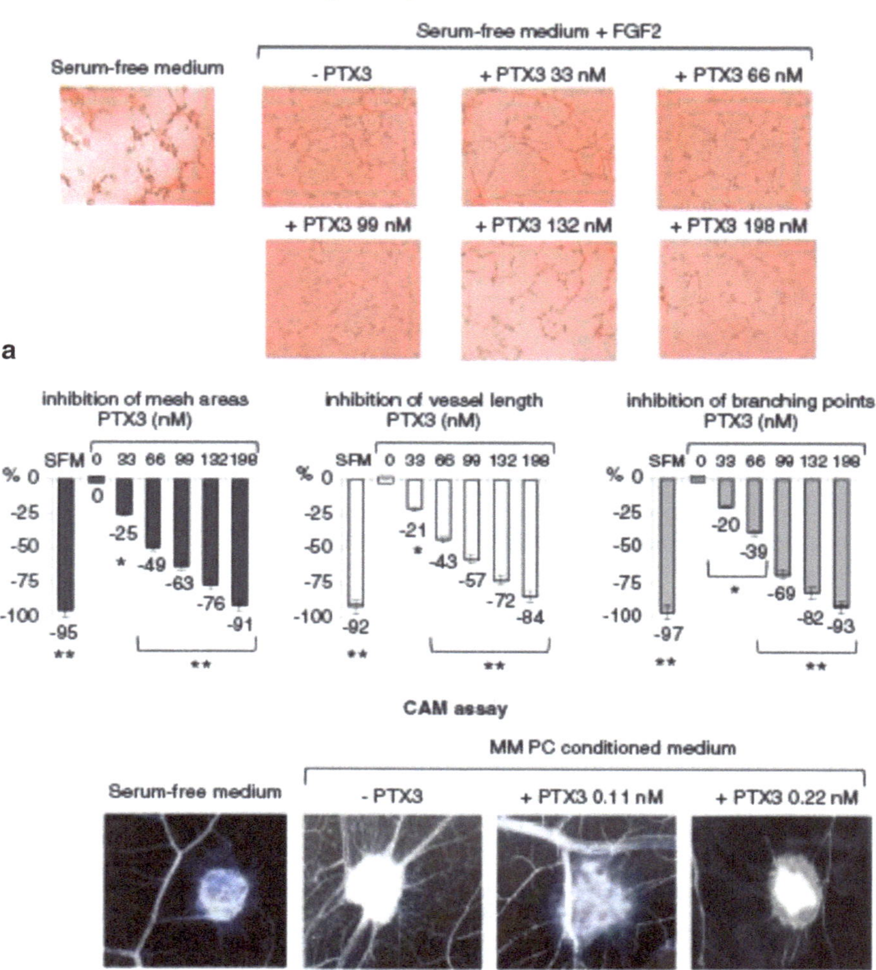

Fig. 2.11 PTX3 inhibits angiogenesis in Matrigel and chick embryo chorioallantoic membrane (CAM) assays (**a**). Multiple myeloma (MM) endothelial cells (ECs) from 18 patients were seeded on Matrigel in serum-free medium (SFM) and SFM supplemented with FGF2 and PTX3 doses. After an 8 h incubation, inhibition of vessel mesh areas, length, and branching points was examined by image analysis software. Matrigel pictures from a representative patient are shown (**b**). CAMs were treated with sponges loaded with SFM (negative control: physiological angiogenesis) or MM plasma cell (PC) conditioned media from 10 MM patients in the absence or presence of PTX3. Angiogenesis was measured and is expressed as mean ± SD vessel counts. (Reproduced from Basile et al. 2013)

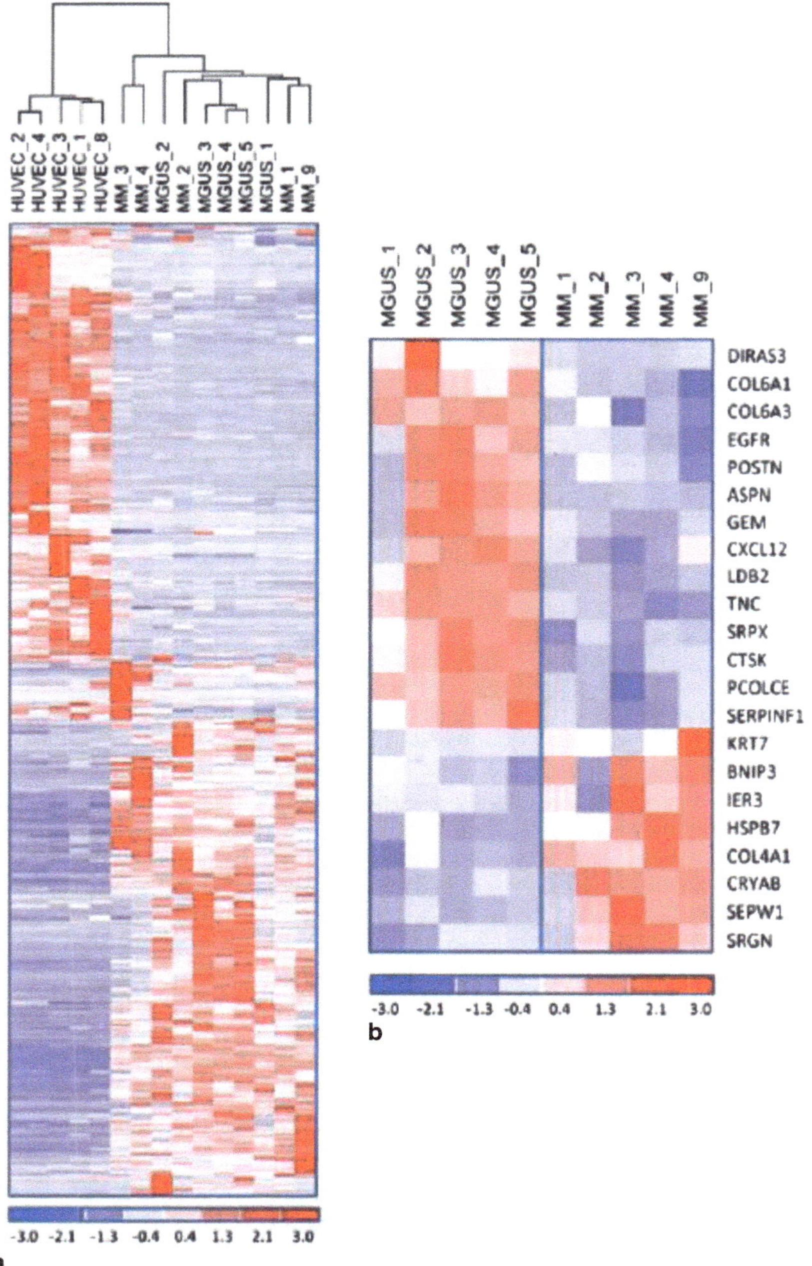

Fig. 2.12 Unsupervised (**a**), and supervised (**b**), analysis of gene expression profiles from a data set composed of five monoclonal gammopathy of undetermined significance endothelial cells, five multiple myeloma endothelial cells, and five HUVEC samples. In (**b**) identification of the 22 genes differentially expressed in five multiple myeloma endothelial cells versus five monoclonal gammopathy of undetermined significance endothelial cells. Color scale bar, the relative gene-expression changes normalized by the SD; the color changes in each row is gene expression relative to the mean across the samples (with gene symbols). (Reproduced from Ria et al. 2009)

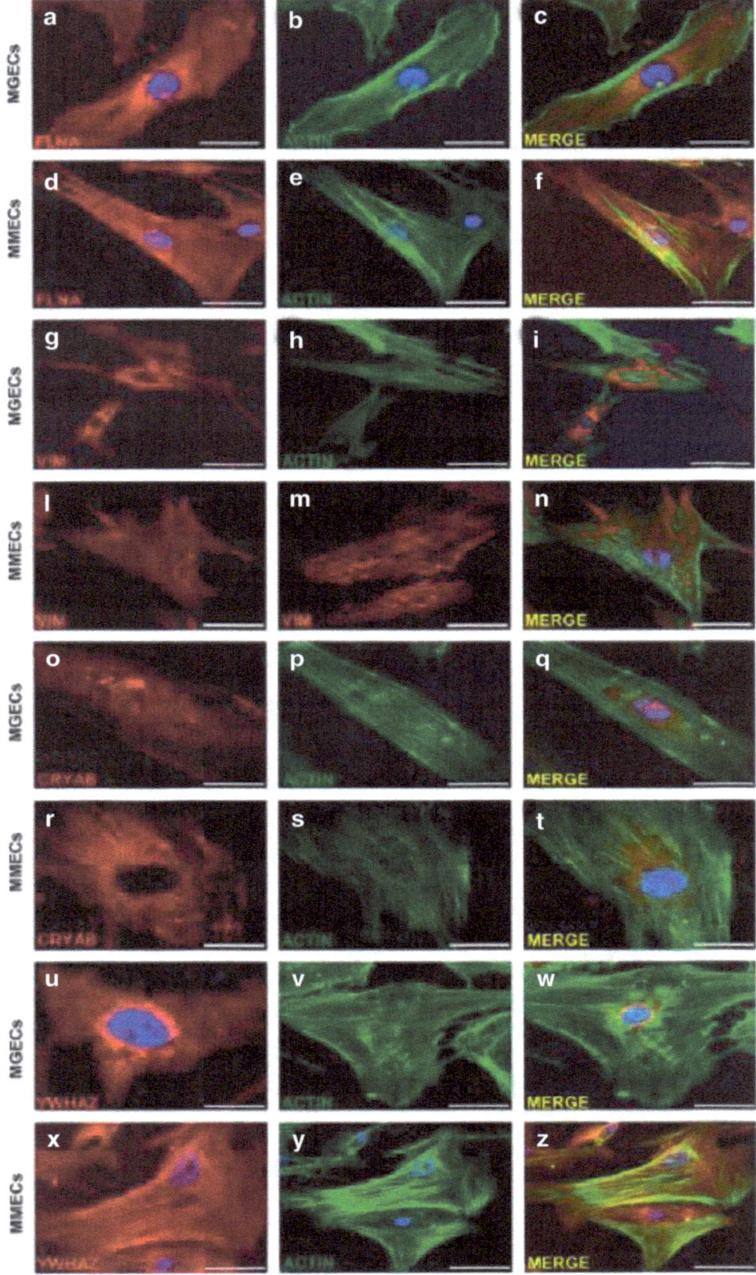

Fig. 2.13 Confocal dual immunofluoresecence of (*red* signal; **a**, **d**) filamin-a (FLNA); (**g**, **l**, **m**) vimentin (VIM); (**o**, **r**) α-crystallin β (CRYAB); (**u**, **z**) YWHAZ; and (*green* signal; **b**, **e**, **h**, **p**, **s**, **v**, **y**) actin filaments in MGECs and MMECs. Merge is co-localization signal with actin. Cell nuclei were stained with TO-PRO-3 iodide (*blue* signal). These four proteins are found overexpressed in MMECs and their expression was enhanced by VEGF, FGF-2, HGF/SC and MM plasma cell culture medium in step with enhancement of MMEC angiogenesis. Their silencing RNA knockdown affected MMEC functions, such as spreading, migration, and tubular morphogenesis. (Reproduced from Berardi et al. 2012)

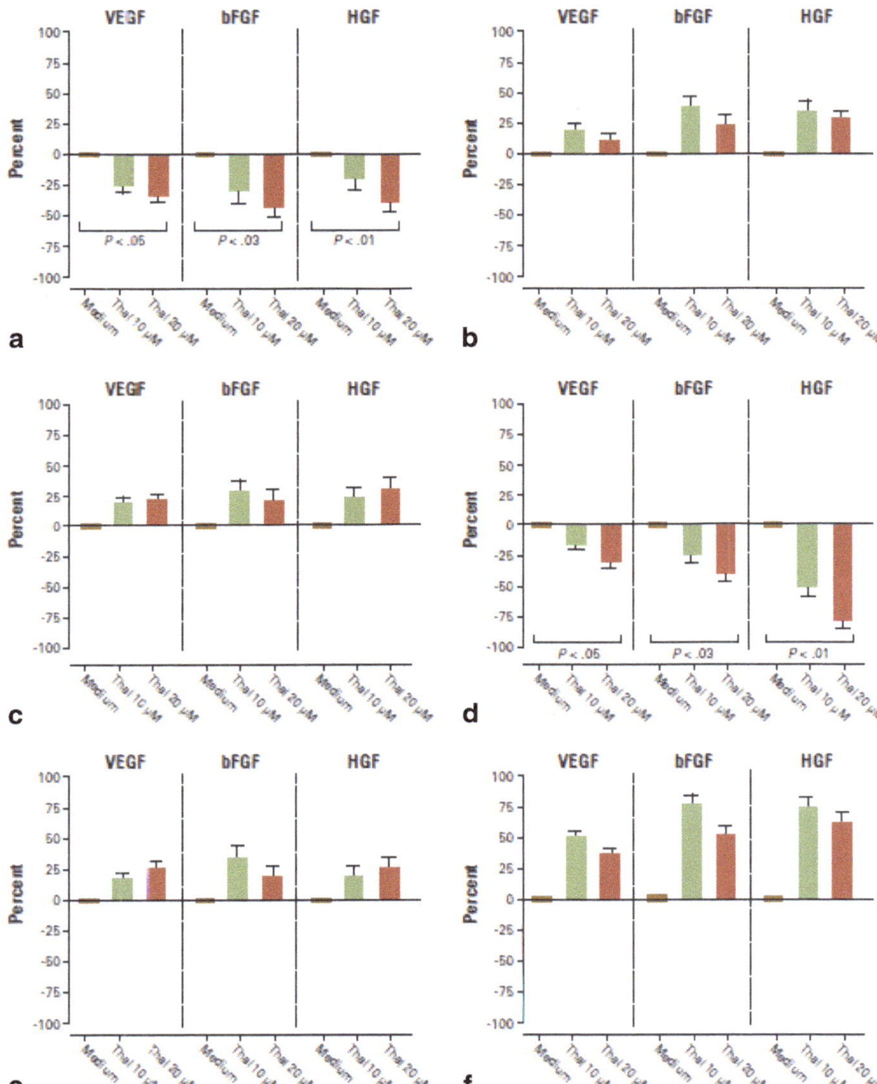

Fig. 2.14 Expression levels of VEGF, bFGF, and HGF genes in the endothelial cell (EC) types without (medium) and with exposure to thalidomide (thal), as evaluated by the RT-PCR (**a**). Active multiple myeloma endothelial cells (relapse) (**b**), nonactive multiple myeloma endothelial cells(plateau phase) (**c**), monoclonal gammopathy endothelial cells (**d**), Kaposi's sarcoma cell line (**e**), diffuse large B-cell non-Hodgkin's lymphoma endothelial cells (**f**), human umbilical vein endothelial cells (HUVECs). (Reproduced from Vacca et al. 2005)

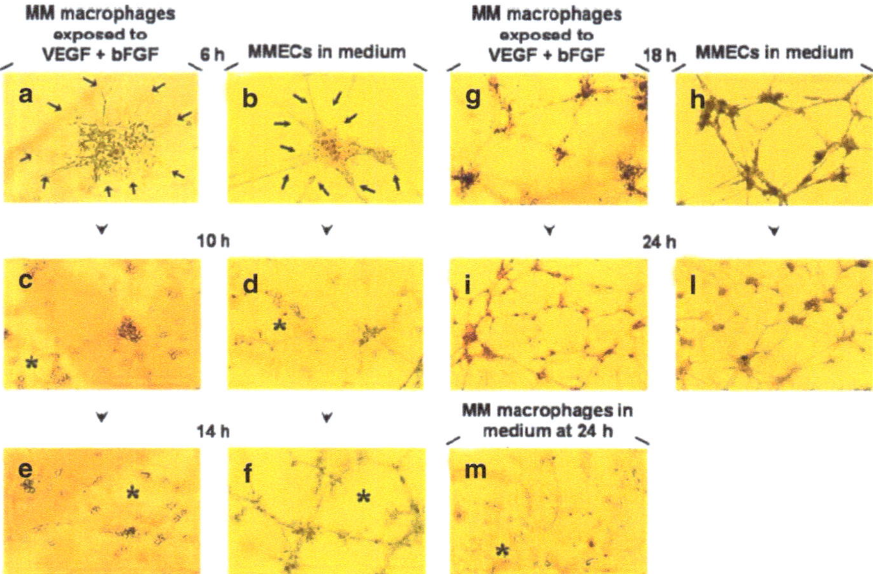

Fig. 2.15 Capillarogenic activity of macrophages and endothelial cells of a patient with active multiple myeloma (MM, relapse). Macrophages seeded on the Matrigel surface in their medium supplemented with vascular endothelial growth factor plus basic fibroblast growth factor (VEGF+bFGF) for 24 h (**a**, **c**, **e**, **g** and **i**): (a) at 6 h cells assemble in nests putting out sprouts (*arrows*); (**c** and **e**) at 10–14 h cells form empty areas (*asterisks*) like the meshes of a grid; (**g** and **i**) at 18–24 h cells form a capillary-like network. A similar capillarogenic activity was displayed at any time by paired multiple myeloma patient-derived endothelial cells (MMECs) cultured in their medium alone (**b**, **d**, **f**, **h** and **l**). Macrophages cultured in their medium alone for 24 h did not produce tube like structures (**m**). (Reproduced from Scavelli et al. 2008)

giogenesis-related functions, including spreading, migration and tubular morphogenesis (Berardi et al. 2012). Thalidomide affects the expression levels of VEGF, FGF-2 and HGF in MM endothelial cells, as evaluated by the RT-PCR (Fig. 2.14; Vacca et al. 2005).

2.7 The Role of Macrophages and Mast Cells

It is well known that among inflammatory cell found in tumors, tumor associated macrophages and mast cells support tumor growth and neovascularization by producing a wide array of angiogenic cytokines. Mast cell- and macrophages-derived growth factors able to promote tumor development and angiogenesis include TNF-α, TGF-β1, FGF-2, VEGF, PDGF, IL-8, osteopontin, and nerve growth factor (NGF). On the contrary, mast cell- and macrophages-produced cytokines that may

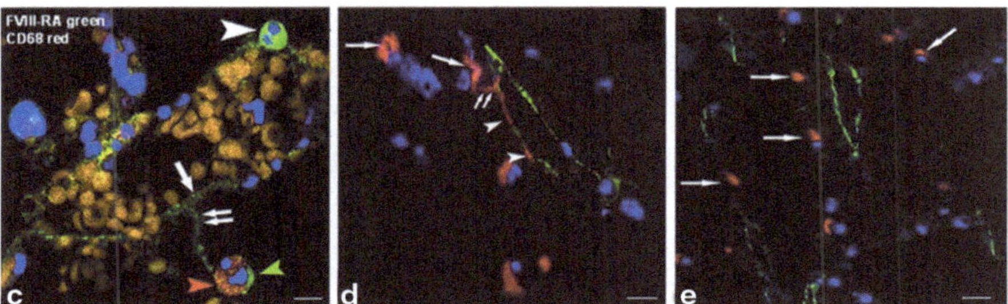

Fig. 2.16 A microvessel lined by flattened FVIII-RA positive multiple myeloma endothelial cells (*arrow*), a FVIII-RA positive macrophage (*arrowhead*) showing protrusions connected to multiple myeloma endothelial cells, and another macrophage containing double-labeled CD68 (*red arrowhead*) and FVIII-RA (*green arrowhead*) granules in the cytoplasm and connected to multiple myeloma endothelial cells by an FVIII-RA positive cytoplasmic protrusion (*double arrow*). Erythrocytes (*orange*) are well recognizable inside the lumen (**a**). Another microvessel formed by FVIII-RA positive (*green*) multiple myeloma endothelial cells and CD68 positive (*red, arrowheads*) tracts that belong to the cytoplasmic protrusions (*double arrow*) of macrophages, some of which are arrowed (**b**). The MGUS microvessels are formed by only FVIII-RA positive endothelial cells: macrophages (*arrows*) are randomly scattered in the tissue and independent of them (**c**). (Reproduced from Scavelli et al. 2008)

participate in anti-tumor response include IL-1, IL-2, IL-4, IL-10, and interferon gamma (IFN-γ) (Ribatti 2013).

When bone marrow macrophages from MM patients are exposed to VEGF and FGF-2, they transform into cells functionally and phenotypically similar to paired MM endothelial cells, and generate capillary-like networks mimicking those of MM endothelial cells (Fig. 2.15; Scavelli et al. 2008). By contrast, macrophages from nonactive MM, MGUS, and benign anemia patients display similar, albeit weaker features. Endothelial cell-like macrophages and apparently typical macrophages contribute sizeably to the formation of the neovessel wall in patients with active MM, whereas their vascular supply is minimal in nonactive MM, and absent in MGUS patients and control patients (Scavelli et al. 2008). In patients with active MM, FACS analyses on freshly isolated bone marrow mononuclear cells revealed higher percentages of CD14/CD68 double-positive cells than in those with nonactive disease and MGUS. Furthermore, in active MM patients bone marrow biopsies displayed macrophages with both endothelial cell-like (i.e. CD68/FVIII-RA double positive) and apparently typical (i.e. CD68 positive/FVIII-RA negative) features located in the microvessel wall and collaborating with MM endothelial cells to line the vessel lumen. Figures of this type were rare in nonactive MM patients and absent in MGUS. Thus, macrophage involvement in the vasculogenic pathway proceeds in step with MM activity, and with progression of plasma cell tumors as well (Fig. 2.16; Scavelli et al. 2008).

Bone marrow angiogenesis and mast cell density counts are highly correlated in patients with nonactive and active MM and in those with MGUS, and that both parameters increase simultaneously in active MM (Ribatti et al. 1999). Ang-1 is a crucial promoter of MM cell growth by stimulating angiogenesis. Experimental

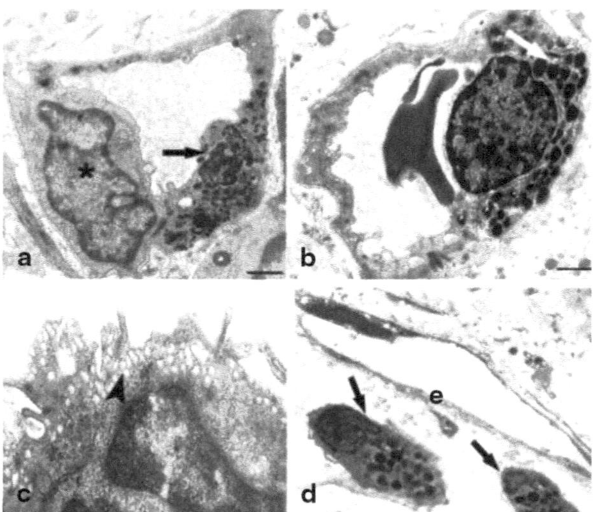

Fig. 2.17 Ultrastructural features of vessels from MM (**a**–**c**), and MGUS (**d**), bone marrow biopsy specimens. In A–C, the vessels are lined by MCs filled with electron-dense granules (A and B, *arrow*) and by thick endothelial cells (A, *asterisk*) characterized by numerous endocytotic vescicles (C, *arrowhead*). In D, a large bone marrow vessel is lined by thin endothelial cells (**e**), surrounded by two mast cells (*arrows*). (Reproduced from Nico et al. 2008)

evidence indicates that Ang-1 secreted by primary murine mast cells promote marked neovascularization in an *in vivo* transplantation assay (Nakayama et al. 2004). These authors demonstrated that primary mast cells accelerate tumor growth by established plasmocytoma cell lines, while Ang-1-neutralizing antibodies significantly reduced the growth of plasmocytomas containing mast cells.

At the ultrastructural level vessels from MM biopsies are lined by mast cells whose cytoplasms is filled with numerous and irregularly shaped electron dense granules (Fig. 2.17; Nico et al. 2008). Moreover, thick endothelial cells, containing endocytotic vescicles, but lacking granules, are connected by a junctional system with the mast cells lining the vessel wall, whereas the vessels from MGUS biopsies are lined with thin endothelial cells, often surrounded by mast cells (Nico et al. 2008).

These ultrastructural findings have been confirmed by confocal laser microscopy using double anti-tryptase (to mark mast cells) and anti-FVIII-RA (to mark endothelial cells) antibodies. Vessels from MM biopsies displayed regions stained by FVIII-RA alternating with regions stained by both tryptase and FVIII-RA. In the MGUS biopsies, the vessels were uniformly stained by the anti-FVIII-RA antibody only, while tryptase-positive mast cells were only recognizable perivascularly (Fig. 2.18; Nico et al. 2008).

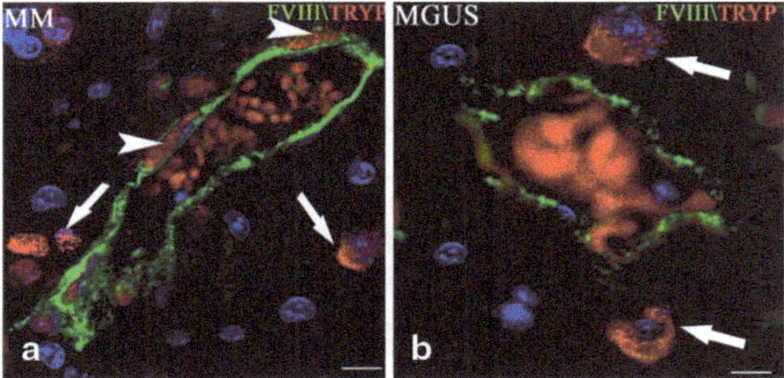

Fig. 2.18 Double FVIII-RA (green) and tryptase (*red*) confocal laser microscopy from multiple myeloma (**a**), and MGUS (**b**), bone marrow biopsy specimens. In A, a multiple myeloma vessel is lined by both endothelial cells positive for FVIII-RA and by mast cellspositive for tryptase (*arrowheads*). Mast cells containing tryptase-positive granules (*arrows*) are also recognizable on the abluminal side of the vessel. In B, a MGUS vessel is lined only by endothelial cells positive for FVIII-RA and is surrounded by tryptase-positive mast cells (*arrows*). (Reproduced from Nico et al. 2008)

2.8 The Role of Endothelial Precursor Cells and of Hematopoietic Stem and Progenitor Cells

A higher number of EPCs has been demonstrated in the bone marrow of patients with active MM than in those with treated MM or with MGUS or in control subjects (Dominici et al. 2001), which reflects the increased angiogenic potential in MM. Zhang et al. (2005) have assessed the view that EPC level and function are correlated with MM activity.

VEGFR-2 (as mRNA and protein) of EPCs was also found to be higher in patients with MM than in healthy subjects, which suggests that it plays a critical role in the emergence and mobilization of EPCs in MM. Frequency of EPCs was responsive to effective treatment with thalidomide and positively correlated with levels of serum M protein and β2-microglobulin, which are indicators of MM activity (Zhang et al. 2005).

Ria et al. (2008a) demonstrated that hematopoietic stem and progenitor cells of MM patients are able to differentiate into cells with endothelial phenotype (Figs. 2.19 and 2.20). Hematopoietic stem and progenitor cells gradually lost CD133 expression and acquired VEGFR-2, factor VIII RA, and VE—cadherin expression, and became capillarogenic in the Matrigel assay (Ria et al. 2008a).

2.9 Prognostic Value of Angiogenesis in Multiple Myeloma

The close correlation between bone marrow microvascular density and LI % in MM was confirmed by Rajkumar et al. (2000), and angiogenesis is a prognostic factor for survival in MM (Sezer et al. 2000). Serum FGF-2, VEGF and

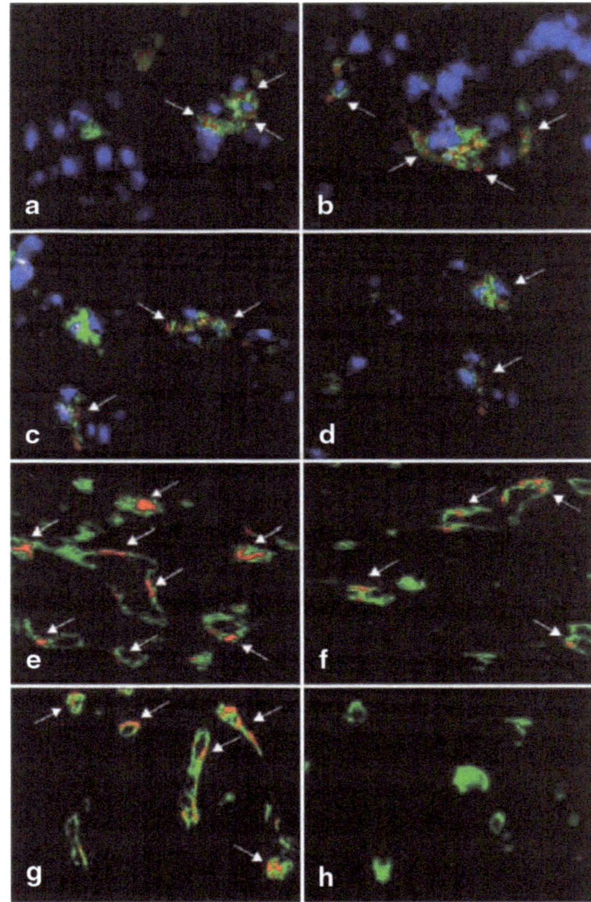

Fig. 2.19 Confocal microscopy analysis of CD133 and endothelial cell markers on bone marrow endothelial cells and neovessels. Some FVIII-RA positive (**a**),VEGFR-2 positive (**b**),VE-cadherin positive (**c**), and Tie2/Tek positive (**d**), clustered multiple myeloma endothelial cells (*green* stained) coexpress CD133 (*red* stained; *arrows*). CD133 positive hematopoietic precursor stem cells are shown in the neovessel wall together with FVIII-RA positive (**e**),VEGFR-2 positive (**f**), and VE-cadherin positive (**g**), multiple myeloma endothelial cell (*arrows*). (**h**), MGUS patient with no CD133 expression inside the FVIII-RA positive vessels. (Reproduced from Ria et al. 2008a)

HGF/SC decrease significantly after successful therapy in MM, whereas little or no reduction is apparent in nonresponders, so that these factors could be of prognostic relevance (Sezer et al. 2000). Accordingly, lowering of bone marrow MVD parallels response to high-dose chemotherapy with autologous stem cell transplantation, which corresponds to a significantly longer progression-free survival compared to persistent high MVD and nonresponse (Sezer et al. 2001). Moreover, bone marrow angiogenesis increases with MM progression from Durie & Salmon stage I to II–III. Its correlation with plasmocytosis and serum β2-microglobulin suggests that MM expansion needs bone marrow vascularization, which could thus be a prognostic factor (Niemoller et al. 2003).

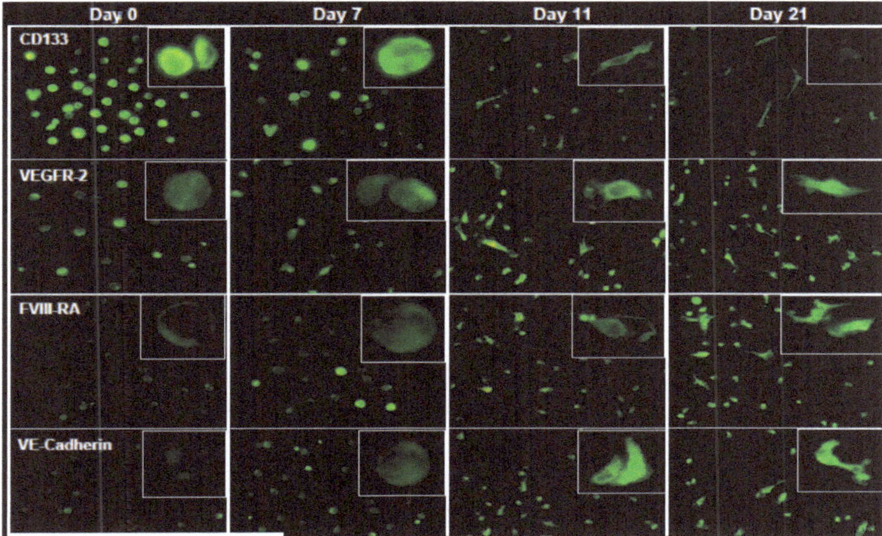

Fig. 2.20 Immunofluorescence analysis of CD133 and endothelial cell markers during hematopoietic stem precursor cells differentiation in a representative patient. Note the gradual loss of CD133 and the parallel increase of VEGFR-2, FVIII-RA, and VE-cadherin on hematopietic stem precursor cells during their differentiation culture. Insets, changes in hematopoietic stem precursor cells shape. (Reproduced from Ria et al. 2008a)

A study of MVD in plasmocytoma and paired bone marrow biopsies was made by Kumar et al. (2003b). They showed 'high-grade' angiogenesis in 64 % of plasmocytomas, whereas angiogenesis was absent in all bone marrows. Patients with high MVD were more likely to progress to MM and had a shorter progression-free survival compared to those with 'low-grade' angiogenesis. Bone marrow MVD at the time of initial diagnosis has a prognostic value for overall and progression-free survival in patients undergoing autologous transplantation as a frontline therapy (Kumar et al 2004).

Chapter 3
Angiogenesis in Lymphomas

3.1 General Features of Lymphomas

Lymphomas constitute a large group of more than 40 lymphoproliferative disorders, classified on the basis of morphological, immunological, genetic, and clinical criteria. The importance of angiogenesis in lymphoproliferative disorders has been studied in relation to their impact on the prognosis of patients, suggesting high relevance in different types of lymphomas (Kini 2004; Koster and Raemaekers 2005; Ruan et al. 2009).

Non Hodgkin's lymphomas (NHL) are a heterogeneous group of lymphoproliferative malignancies with different patterns of behavior responses to treatment (Table 3.1). B-cell lymphomas represent approximately 88 %, and T- and NK-cell lymphomas 12 % respectively, of all NHL. Among B-cell lymphomas, the incidence of diffuse large B-cell lymphomas (DLBCL) is 30 %, of follicular lymphoma (FL) 25 %, of extranodal marginal zone lymphoma of mucosa-associated lymphatic tissue mucosa associated lymphoma tissue (MALT lymphoma) 7 %, of chronic lymphocytic leukemia 7 %, of mantle cell lymphoma 5 %.

Lymphoid tumors are generally divided into one of two categories, namely indolent lymphomas versus aggressive lymphomas, based upon on the characteristics of the disease at the time of presentation and the patients' life expectancy if the disease is left untreated. Generally, T-cell lymphomas have a more aggressive clinical behavior than B-cell lymphomas of comparable histology and patients with mantle cell lymphomas or anaplastic large lymphomas have a 5-year survival rate of approximately 30 and 80 %, respectively (Fisher et al. 1998).

3.2 In Vitro and Vivo Experimental Models

Conditioned media of lymphomas cells induced a five-fold increased proliferation of cultured endothelial cells, suggesting the release of a soluble pro-angiogenic factor (Brandvold et al. 2000). Human lymphoid tumor cells constitutively produce significant amounts of the ECM degrading enzymes MMP-2 and MMP-9

D. Ribatti, *Angiogenesis and Anti-Angiogenesis in Hematological Malignancies*,
DOI 10.1007/978-94-017-8035-3_3, © Springer Science+Business Media Dordrecht 2014

Table 3.1 The Working Formulation for clinical usage classification of non-Hodgkin's lymphomas

Low-grade
Malignant-lymphoma, small lymphocytic
Plasmacytoid
Malignant-lymphoma, follicular, predominantly small cleaved cell
Malignant-lymphoma, follicular, mixed small cleaved and large cell
Intermediate-grade
Malignant lymphoma, follicular, predominantly large cell
Malignant lymphoma, diffuse, small cleaved cell
Malignant lymphoma, diffuse, mixed small and large cell
Malignant lymphoma, diffuse, large cell
Cleaved
Non-cleaved cell
High-grade
Malignant lymphoma, large cell, immunoblastic
Plasmacytoid
Clear cell
Polymorphous
Malignant lymphoma, lymphoblastic
Convoluted
Nonconvoluted
Malignant lymphoma, small non-cleaved cell
Burkitt's
Non-Burkitt's

(MMP-2 and MMP-9), as demonstrated by sodium dodecyl sulphate polyAcrylamidegel electrophoresis (SDS-PAGE) gelatin zymography and *in situ* hybridization (Vacca et al. 2000). Moreover, human lymphoid tumor cells are able to interact with ECM components vitronectin and fibronectin and this interaction it is mediated by $\alpha v\beta 3$ integrin, allowing them to adhere to the substratum and enhancing their proliferation and protease secretion (Vacca et al. 2001b).

Lymphomas cells are able to induce an angiogenic response when tested *in vivo* in the hamster check pouch model (Wolf and Hubler 1975). Similarly, lymphoma bioptic specimens, when implanted on the chick embryo CAM evoked a strong angiogenic response (Ribatti et al. 1990). The angiogenic response did not correlate with either the malignancy grade or the immunologic phenotype of the tumors. Different human Burkitt's lymphoma cells when are inoculated onto the CAM formed solid tumors (Becker et al. 2012). However, Epstein-Barr virus positive cells induced massive recruitment of chick leukocytes at the tumor border and the development of granulation tissue with large number of blood and lymphatic vessels, although all cell lines tested have almost identical VEGF and VEGFR expression (Becker et al. 2012). Finally, conditioned medium from FL and DLBCL samples induced a strong angiogenic response in the CAM assay (Cocco et al. 2012).

3.3 Angiogenesis in Normal Lymph Nodes and in Benign Lymphadenopathies

The lymph node microvasculature consists of arterioles, metarterioles, anastomosing capillaries, small venules and high endothelial post-capillaries venules and small veins. Dense plexuses of capillaries arise from arterioles in the medullary cords, in the periphery of the deep cortex units, and in the outermost stratum of the extrafollicular zone of peripheral cortex. In contrast, the folliculo-nodules and center of the deep cortex units are little vascularized by a loose capillary network, while no vessels occur in the subsinus layer (Davidson et al. 1973; Anderson and Anderson 1975; Herman 1980). When tissue fragments from normal lymph nodes are grafted *in vivo* on the chick embryo CAM, stereomicroscopic observation of the area around the implant revealed little hyperemia a small number of growing vessels (Ribatti et al. 1990).

In both reactive lymph nodes and in lymph nodes with follicular lymphomas, MVD is higher in the paracortex than in the follicles and that there are no difference in MVD between reactive germinal centers and neoplastic follicles (Passalidou et al. 2003). Moreover, MVD in the paracortex in reactive lymph nodes is higher than in diffuse large large lymphomas (Passalidou et al. 2003). In FL, several studies have recognized an increase in MVD in reactive parts of affected lymph nodes outside the follicles, as compared to the neoplastic follicles (Koster et al. 2005; Vacca et al. 1999c; Ridell and Norrby 2001; Arias and Soares 2000). Other Authors (Ribatti et al. 1996) have shown that MVD is higher in lymphomas than in reactive lymph nodes and in aggressive than indolent lymphomas, or that MVD in reactive lymph nodes is comparable to that observed in lymphomas (Arias and Soares 2000).

3.4 Angiogenesis in Non Hodgkin's Lymphomas

As concerns the morphological features of tumor blood vessels, two patterns of laminin and type IV collagen expression are recognizable in the perivascular stroma of B cell-NHL, classified accordingly to the Working formulation in low-, intermediate- and high-grade tumors (Ribatti et al. 1996). A granular, speckled and low intensity staining was expressed by laminin, and more frequently associated with the intermediate- and high-grade tumors. A linear continuous staining was co-expressed by laminin and type IV collagen, and was more frequently associated with low-grade tumors. The granular and linear patterns may correspond to different steps in the basement membrane deposition. The granular pattern to the first one, when endothelial proliferation takes place; the linear pattern to the second stage, when basement membrane is completely differentiated (Paku and Paweletz 1991). This hypothesis is in agreement with the evidence that endothelial sprouting is associated with a higher concentration of laminin that type IV collagen around the outgrowing capillaries (Ausprunk and Folkman 1977).

Moreover, the expression of tenascin in the stroma of B-NHL has been investigated and related to histologic malignancy and angiogenesis (Vacca et al. 1996). It is well known that tenascin stimulates angiogenesis and since it forms a long reticulum with long extensions directly from vessels, it could also provide a pathway that favors migration of endothelial cells (Rettig et al. 1994). The presence of an increased number of immature vessels in DLBCL compared with FL, classified accordingly to the World Health Organization (WHO) classification, has been also demonstrated (Passalidou et al. 2003).

At ultrastructural level, the presence of immature capillaries in the stroma of diffuse intermediate-grade, and high-grade B cell-NHL has been shown (Ribatti et al. 1996). These capillaries lack the basement membrane and generally consisted of two endothelial cells arranged in parallel, with thickened cytoplasm, resulting in slit-like lumen. On the contrary, in follicular intermediate-grade and low-grade B cell-NHL, differentiated fenestrated capillaries surrounded by a continuous basement membrane were recognizable. Moreover, a morphological heterogeneity of tumor blood vessels between histologic subtypes of lymphomas has been shown together with different patterns of neovascularization in both low and high-grade B cell-NHL (Crivellato et al. 2003a, b). In low-grade the vessel lumen is formed either by endothelial cell body curving or, more frequently, by the fusion of intracellular vacuoles in poorly differentiated endothelial cells. In high-grade, on the other hand, the prevalent neo-angiogenic pattern is the formation of a slit-like lumen (Crivellato et al. 2003a, b). Both low- and high-grade tumors exhibited development of transluminal bridges, expression of intussusceptive microvascular growth, an alternative mode of tumor vessel growth (Ribatti and Djonov 2012).

As concerns the evaluation of MVD in bioptic specimens, a correlation between microvascular density and histologic subtype in NHL has not been established (Hazar et al. 2003), or differences in microvascular density in patients with chemotherapy-resistant DLBCL and those with chemosensitive lymphomas has not been found (Bairey et al. 2000). Other studied in NHL and in DLBCL, found no correlation between MVD and VEGF expression (Jørgensen et al. 2007; Ganjoo et al. 2006; Ruan et al. 2006). On the contrary, an increased vascularity pre-treatment predicted favorable outcome in terms of progression-free and overall survival in patients in FL patients who received chemotherapy in association with IFN-α2b (Koster et al. 2005), or in FL a high MVD predicted progressive disease and overall survival and correlated with transformation to DLBCL (Jørgensen et al. 2007).

Microvascular density is highest in aggressive subtypes including Burkitt's lymphoma and peripheral T cell lymphoma (PTCL), compared with intermediate in DLBCL and lower in indolent FL (Ribatti et al. 1996). In DLBCL the average MVD correlate with the intensity of VEGF tumor cells immunoreactivity (Gratzinger et al. 2007). On the contrary, another study of the same group on patients affected by DLBCL treated with anthracycline-based chemotherapy showed no correlation between increased MVD and lymphoma cell VEGF expression (Gratzinger et al. 2007). In cutaneous T cell and B cell lymphoma, MVD is higher than in normal skin of a benign cutaneous lymphoproliferative disorder (Vacca et al. 1997; Schaerer et al. 2000; Mazur et al. 2004).

The clinical significance of increased MVD is not clear and difficult to establish because most of the studies different treatment describe heterogeneous populations including a wide variety of histologic subtypes of NHL and regimens. Several studies have demonstrated that high levels of VEGF in lymphoma samples correlate with advanced tumor stage and higher risk for relapsed/refractory disease after standard chemotherapy. In NHL high pretreatment levels of serum VEGF was a prognostic factor for survival in multivariate analysis (Salven et al. 2000). In both T and B cell lymphomas, a negative correlation between the overall survival rate respectively 5-year disease-free survival and the pretreatment serum level of VEGF has been established (Niitsu et al. 2002), while in patients with DLBCL treated with cyclophosphamide, doxorubicin, vincristine, and prednisolone (CHOP), high serum level of VEGF was associated with adverse outcome, having lower values in survivors than in non-survivors (Aref et al. 2004).

VEGF expression was also demonstrated in PTCL, DLBCL, mantle cell lymphoma (MCL), primary effusion lymphoma, chronic lymphocytic leukemia/small lymphocytic lymphoma (CLL/SLL) (Doussis-Anagnostopuolou et al. 2002; Foss et al. 1997; Chen et al. 2000; Kay et al. 2002). An adverse outcome associated with an increased VEGF tissue expression in aggressive and indolent lymphomas of B cell and T cell origin (Kuramoto et al. 2002) has been demonstrated. In angioimmunoblastic T cell lymphoma VEGF-A gene is overexpressed in both tumor and endothelial cells in comparison with reactive lymph nodes in association with a short survival time (Zhao et al. 2004), and an high expression of VEGF-A in aggressive T cell lymphomas as compared to indolent B cell lymphomas has been found (Foss et al. 1997). In contrast, only a minority of indolent FL show variable expression of VEGF-A (Koster et al. 2005; Jørgensen et al. 2007), and transformation from indolent B cell lymphoma to aggressive DLBCL and poor prognostic subgroups within DLBCL are associated with increased VEGF expression (Shipp et al. 2002).

In primary diffuse central nervous system lymphomas (PCNSL), VEGF expression is correlated to MVD and VEGF expression is associated with a longer survival and blood-brain barrier alteration (Takeuchi et al. 2007). In 24 human diffuse large B-cell PCNSL studied by means of immunocytochemistry and confocal laser microscopy has been demonstrated that: (a) Aquaporin 4 (AQP4) expression was directly correlated with Ki-67 index, while AQP4 expression was low in tumor areas with a low Ki-67 index. (b) Different cells participated to vessels formation: $CD20^+$ tumor cells and factor VIII^+ endothelial cells; $AQP4^+$ tumor cells and $CD31^+$ endothelial cells; $CD20^+$ and $AQP4^+$ tumor cells; glial fibrillary acidic protein $(GFAP)^+$ endothelial cells surrounded by $GFAP^+$ tumor cells (Figs. 3.1, 3.2, 3.3, 3.4). Overall, these data suggest the importance of AQP4 in PCNSL due to its involvement in pathogenesis and resolution of cerebral edema. AQP4 is also involved in migration of tumor cells. It was also documented that tumor microvasculature in PCNSL is extremely heterogeneous, confirming the importance of neoangiogenesis in their pathogenesis (Nico et al. 2012).

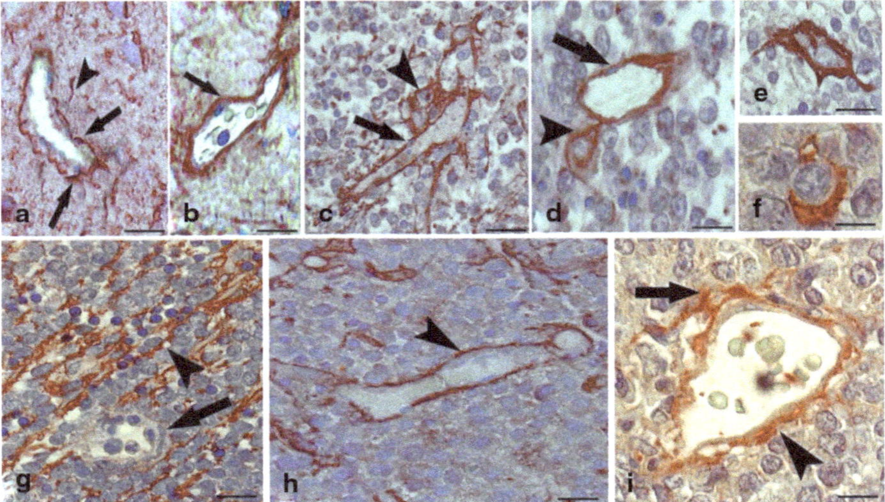

Fig. 3.1 Aquaporin-4 (AQP4) immunocytochemical expression in internal healthy (**a**) and in control (**b**) brain tissues and in primary central nervous system lymphoma (PCNSL) with high (**c**–**g**) and low (**h**–**i**) Ki-67 index. **a**, **b** Control brain sections showing a continuous perivascular AQP4 staining of glial endfeet surrounding (*arrow*) and touching (**a**, *arrowhead*) the vessel wall. **c**–**g** Tumor brain sections with high Ki-67 index showing AQP4-labelled endothelial cells (**c**, **d**, *arrow*), and clusters of labelled tumor cells close to the vessels (**c**, **d**, *arrowhead*), and expressing a cytoplasmic AQP4 staining (**e**, **f**). AQP4 staining (**g**, *arrowhead*) is detectable between rows of tumor cells including an unlabelled vessel (**g**, *arrow*). **h**, **i** Tumor brain sections with low Ki-67 index showing a restored AQP4 perivascular labelling (**h**, **i**, *arrowhead*). A labelled AQP4 tumor cell (**i**, *arrow*) close to a vessel wall is recognizable. (Reproduced from Nico et al. 2012)

The degree of VEGF expression correlated with the expression level of VEGFR-1 and -2 in DLBCL lymphoma cells (Gratzinger et al. 2007), and VEGFR-1, -2, and -3 are expressed in CLL, suggesting the possibility that VEGF acts as an autocrine/paracrine factor (Bairey et al. 2004). Moreover, VEGF prevents apoptosis, and increase phosphorylation of VEGFR-1 and -2, further supporting the existence of an autocrine pro-survival loop in CLL (Lee et al. 2004). Blocking of VEGF and VEGFRs, by using neutralizing antibodies or tyrosine kinase inhibitors, resulted in decreased levels of p-STAT-3 and apoptosis of CLL cells (Lee et al. 2005). High VEGF and VEGFR-1 expression identified a subgroup of patients affected by DLBCL with improved overall survival and progression-free survival when treated with anthracycline-based chemotherapy, suggesting that the autocrine signaling via VEGFR-1 may be susceptible to this therapeutic approach (Gratzinger et al. 2008). When immunodeficient mice engrafted with human DLBCL were treated with antibodies against human or murine VEGFR-1 or VEGFR-2, a significant tumor reduction of 50 % was observed after treatment with human anti-VEGFR-1, but not with murine anti-VEGFR-1. By contrast, inhibition of murine VEGFR-2 resulted in a similar tumor reduction, but inhibition of human VEGFR-2 had no antitumor effect (Wang et al. 2004). Anti-VEGFR-2 antibody was as effective as rituximab,

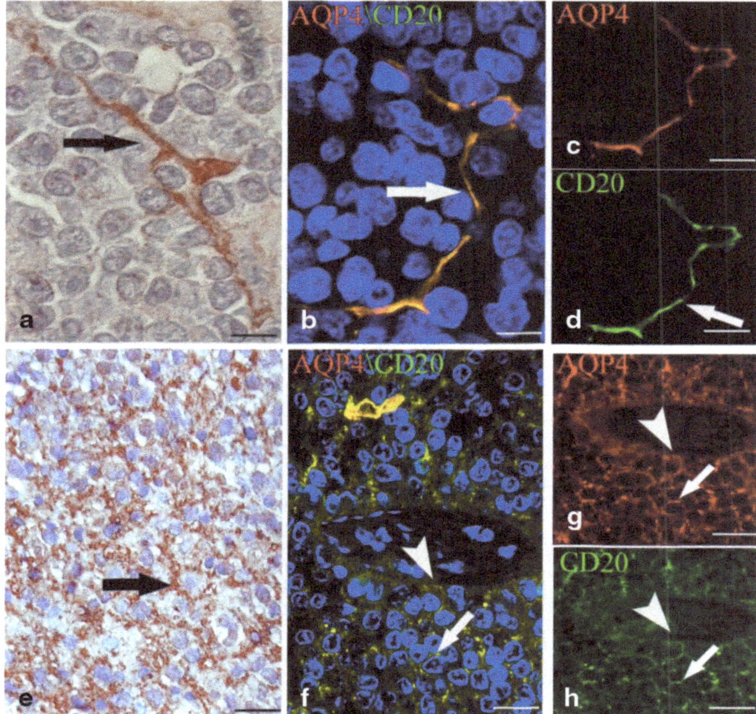

Fig. 3.2 AQP4 immunohistochemistry (**a**, **e**) and AQP4 (in red)/CD20 (in green) (**b**–**d**, **f**–**h**) confocal dual immunofluorescence. Elongated (**a**) or rounded (**e**) tumor cells expressing AQP4 staining in their cytoplasm (**a**, *arrow*) and plasmamembrane (**e**, *arrow*), are labelled by both anti-CD20 (**d**, **h**, *arrow*) and anti-AQP4 antibodies, with a merge fluorescence signal of colocalization on a single cell (**b**, **f**, *arrow*). A vessel lined by tumor cells with a fluorescent signal expression of CD20 (**h**, *arrowhead*) and AQP4 (**g**, *arrowhead*) colocalisation is shown in **f** (*arrowhead*). (Reproduced from Nico et al. 2012)

and when were combined, tumor volume were reduced even more to 75 % (Wang et al. 2004). A less expression of HIF-1, HIF-2 and VEGF in indolent lymphomas, consisting mainly of FL, than in aggressive lymphomas has been observed (Stewart et al. 2002). Accordingly, only a minority of indolent lymphomas, showing histologic transformation to aggressive lymphoma, expressed VEGF-A in contrast to aggressive lymphomas (Ho et al. 2002).

Inhibition of autocrine or paracrine VEGFRs-mediated loops with receptor-specific antibodies suppress the growth of lymphomas by increasing tumor apoptosis and decreasing vascularization, respectively. These results confirm the role of VEGF in lymphomagenesis and support the targeting of VEGFRs as a therapeutic approaches for aggressive lymphomas.

Other angiogenic growth factors may contribute to the angiogenic process and tumor progression in NHL. Among these, FGF-2 is one of the best characterized of

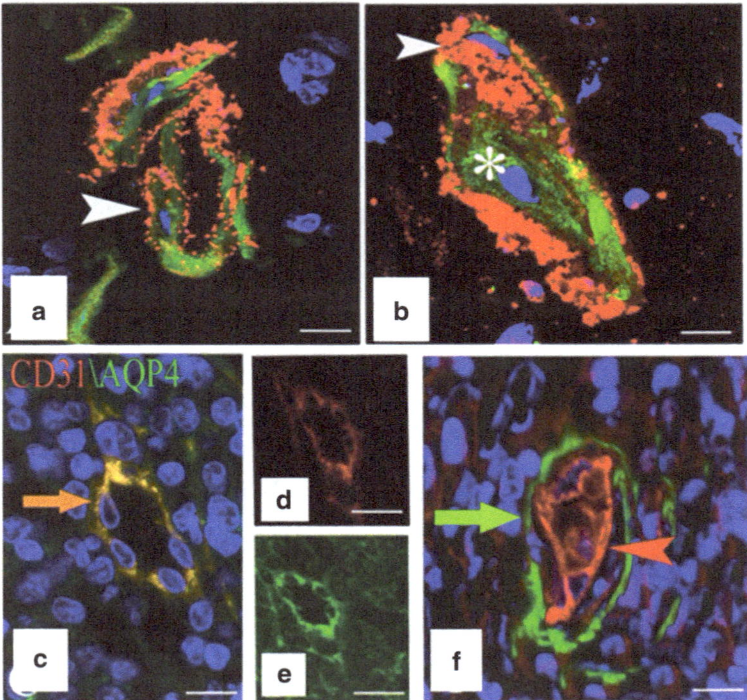

Fig. 3.3 **a**, **b** Confocal dual immunofluorescence reaction of FVIII (in red)/CD20 (in green) in tumor sections. The vessels are lined by FVIII fluorescence endothelial cells (**a**, *arrowhead*) and CD20 labelled tumor cells (**b**, *asterisk*). **c**–**f** Confocal dual immunofluorescence reaction of CD31 (in red)/AQP4 (in green) in tumor (**c**, **e**) and control (**f**) sections. A merge orange signal of AQP4/CD31 colocalization is detectable in the tumor vessel on a single endothelial cell (**c**, *arrow*). A separate endothelial CD31 (**f**, *arrowhead*) and glial AQP4 (**f**, *arrow*) fluorescence signal is detectable in control tissue (**f**). (Reproduced from Nico et al. 2012)

the pro-angiogenic cytokines. Because of its pleiotropic activity that may affect both tumor vasculature and tumor parenchyma, FGF-2 may contribute to cancer progression by inducing neovascularization, as well as by acting directly on tumor cells.

Various lymphoblastoid cell lines secrete FGF-2 (Vacca et al. 1998). Pazgal et al. (2002) measured FGF-2 serum concentration in patients with NHL before and after treatment, conducted an immunohistochemical study to determine the expression of FGF-2 and FGFR-1 and MVD and evaluated the prognostic significance of FGF-2 and FGF receptor-1 (FGFR-1 expression). They demonstrated that FGF-2 expression was correlated with poor survival and progression-free survival, while FGFR-1 expression was correlated with decreased rate of achievement of complete remission. Moreover, they did not detect a significant change in serum FGF-2 levels after two-three cycles of chemotherapy, nor they did find a correlation between microvascular density and NHL histology or grade or between MVD and prognosis. Moreover, in malignant lymphoma high pretreatment levels of FGF-2

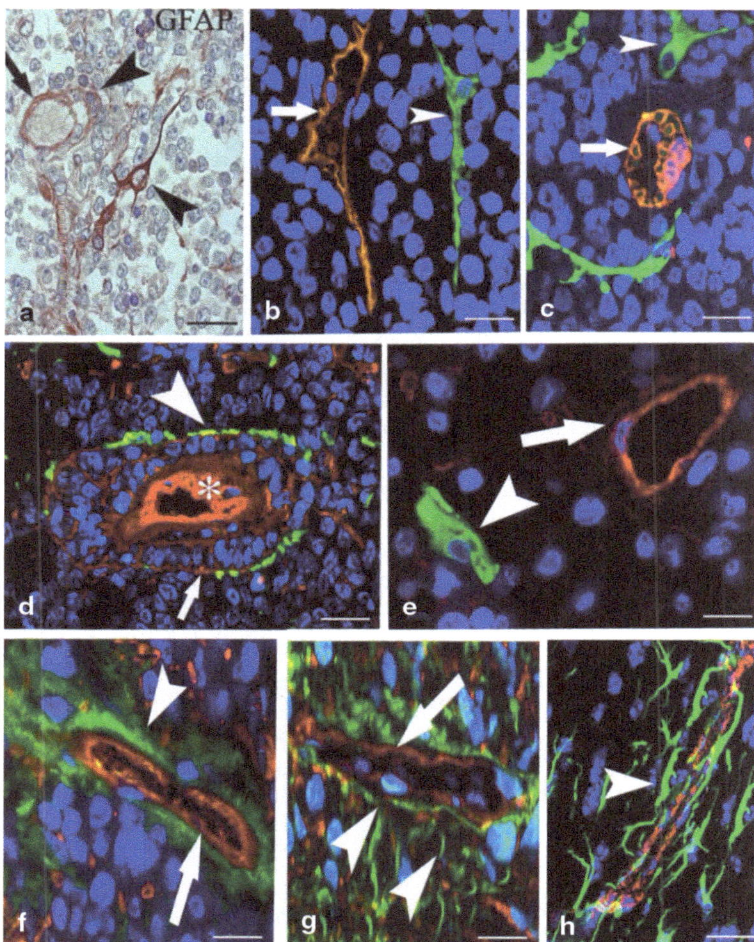

Fig. 3.4 Glial fibrillary acidic protein (GFAP) immunohistochemistry (**a**) and FVIII (in red)/GFAP (in green) (**b**–**g**) confocal dual immunofluorescence. **a** Tumor vessels are lined by GFAP marked endothelial cells (*arrow*) surrounded by GFAP stained tumor cells (*arrowhead*). **b**–**e** Tumor vessels with high Ki-67 index show an orange fluorescence signal expression of FVIII/GFAP colocalization on a single endothelial cell (**b**, **c**, *arrow*) near to GFAP labelled tumor cells (**b**, **c**, *arrowhead*). Tumor vessels with thickened (**d**, *asterisk*) or thin (**e**, *arrow*) endothelium positive to FVIII, surrounded by tumor cells expressing a separate GFAP (**d**, **e**, *arrowhead*) and FVIII fluorescence (**d**, *arrow*), are recognizable. **f**–**h** Tumor vessels with lowKi-67 index (**f**), internal healthy (**g**), and control brain (**h**) tissues showing an endothelial FVIII red fluorescence (*arrow*) are enveloped by a continuous layer of GFAP green fluorescence glial processes (*arrowhead*). (Reproduced from Nico et al. 2012)

was a prognostic factor for survival in multivariate analysis, independently of other risk factors, including serum lactate dehydrogenase and number of extranodal sites (Salven et al. 2000). Soluble VEGF, FGF-2, and PDGF-β levels decline after radiotherapy in NHL, suggesting that may have predictive significance for response to treatment and recurrence (Ria et al. 2008b).

3.5 The Role of Myelo-Monocytic Cells and Circulating Endothelial Cells

At least three categories of pro-angiogenic bone marrow-derived circulating cells have been implicated in tumor angiogenesis: (a) cells that contribute directly to the structural components of angiogenesis, including EPC and pericyte progenitors; (b) myeloid progenitors subsets that can differentiate into endothelial-like cells and incorporate into the tumor neovessels; (c) a large heterogeneous group of cells of monocytic lineage that functions as vascular modulators that are not physically part of the vasculature (Ribatti 2009).

It is well established that neoangiogenesis and growth of murine lymphomas is dependent on the recruitment of bone marrow-derived proangiogenic hematopoietic cells (De Palma et al. 2003). Increased hematopoietic infiltration by myeloid progenitors $CD68^{+}$ and VEGFR-1^{+} and producing VEGF-A has been correlated with histologic subtypes of lymphoma, suggesting their involvement in the development of a pro-angiogenic phenotype (Vacca et al. 1999; Ruan et al. 2006; Farinha et al. 2005). In aggressive subtypes of Burkitt's lymphoma and DLBC, VEGF-A producing $CD68^{+}$ $VEGFR1^{+}$ myelo-monocytic cells are closely associated to new-formed blood vessels (Ruan et al. 2006). Genetic depletion of this subpopulation of $CD68^{+}$ $VEGFR1^{+}$ myelo-monocytic cells was sufficient to inhibit angiogenesis in various tumor experimental models, including lymphoma (De Palma et al. 2003).

3.6 The Role of Macrophages and Mast Cells

Angiogenesis extent and macrophage density increase simultaneously with pathological progression in B cell NHL, suggesting that an increase number of macrophages may be recruited and activated locally by malignant B cells, and that angiogenesis associated with B cell-NHL may be induced, at least, partly, by angiogenic factors secreted by macrophages (Vacca et al. 1999c). High number of intratumoral macrophages correlate with poor prognosis in FL treated with chemotherapy alone, and rituximab appears to circumvent the unfavourable prognosis associated with high number of macrophages (Canioni et al. 2008; Taskinen et al. 2007).

The extent of angiogenesis has been correlated with the number of mast cells in B cell-NHL and both counts increase in step with the increase of malignancy grade (Ribatti et al. 1998, 2000; Crivellato et al. 2002). Tryptase together other angiogenic factors stored in mast cell secretory granules may contribute to angiogenesis in B cells-NHL (Ribatti et al. 1998; Crivellato et al. 2002). In an ultrastructural study

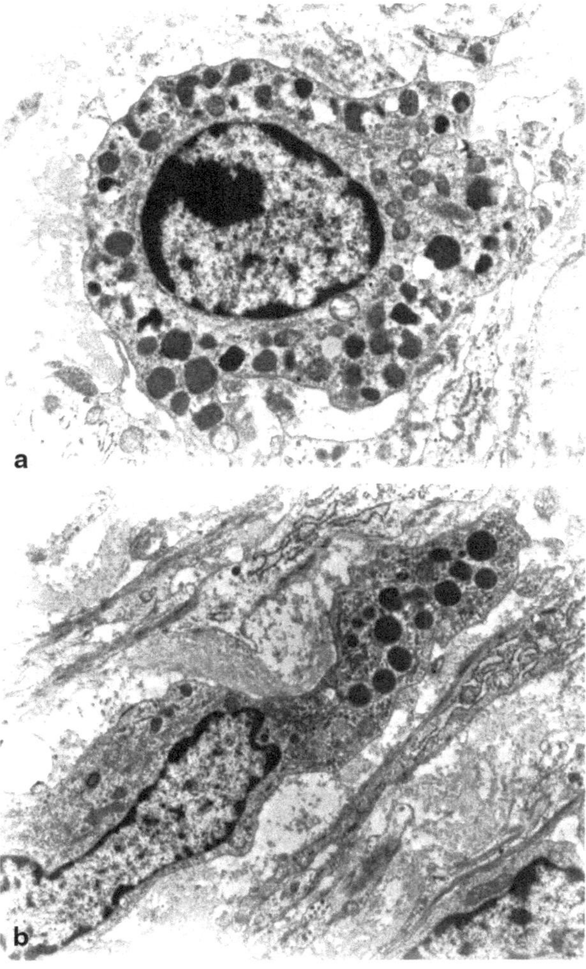

Fig. 3.5 Ultrastructural appearance of mast cells from low-grade B-non Hodgkin lymphoma. In **a** typical mast cell with a large nucleus and irregular surface contour. The cytoplasm is filled with numerous, various shaped and moderately electron-dense secretory granules. Note that some granules show focal losses of matrix material in the form of lucent areas within the dense content of a single granule. In **b** an elongated fibroblast-like mast cell showing a limited granule repertoire with a homogeneous structural pattern (Reproduced from Crivellato et al. 2002)

of samples of B cell NHL has been demonstrated the presence of a heterogeneous population of mast cells characterized by the presence of granules with semilunar aspect and containing scrolls (Fig. 3.5) (Crivellato et al. 2003b). Semilunar granules are the expression of a slow but progressive release of angiogenic factors due to chronic, progressive stimulation of mast cell degranulation, while in the granules containing scrolls is stored tryptase, an angiogenic factor (Fig. 3.6) (Ribatti and Crivellato 2012b). In B cell CLL there is a striking association between the number of mast cells and microvascular density in bone marrow and both increase as the disease progresses (Ribatti et al. 2003b).

Circulating EPCs have been detected within the blood flow during tumor growth and several evidence indicate that bone marrow-derived circulating EPCs contribute to tumor growth and tumor angiogenesis (Ribatti 2007). An increased number of

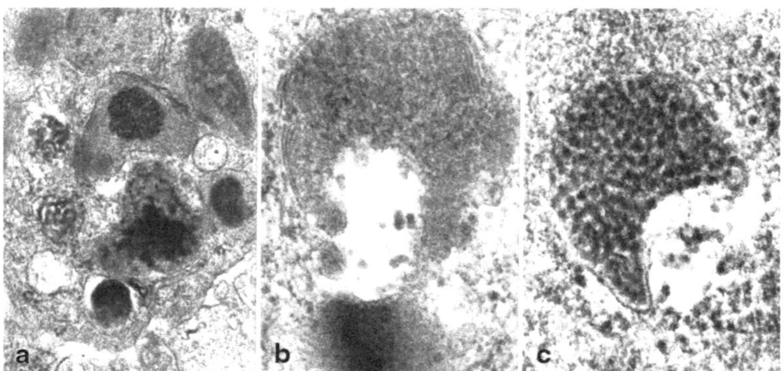

Fig. 3.6 Ultrastructural findings of mast cell granules from high-grade B-NHL. Note in **a** a heterogeneous mixture constituting of coarse particles and scrolls, thick threads and finely granular material or a mixture of this and scrolls. Note in **b** a semilunar granular aspect and **c** a full complement of dense particles recognizable in this type of granule. (Reproduced from Crivellato et al. 2002)

CD133^{+} CD34^{+} VEGFR-2^{+} circulating EPCs has been found in younger patients and those with aggressive NHL, and the levels of circulating EPCs decreased following complete response to treatment (Igreja et al. 2007). Moreover, lymph node EPCs were detected in vascular structure and in the stroma and correlated with an increased angiogenesis in indolent lymphoma (Igreja et al. 2007). In angiogenesis-defective Id-mutant mice, VEGFR-2^{+} EPCs constitute >90 % of tumor vessels following wild-type bone marrow rescue in a murine xenograft model of aggressive B cell lymphoma (Lyden et al. 2001).

Within immunodeficient mice engrafted with lymphoma cells, a significant increase in number of circulating endothelial cells (CECs) was observed after a period of 21 days, they being correlated with tumor size and serum level of VEGF (Monestiroli et al. 2001). An increase in CECs in patients with lymphoma compared with to the control cases has been reported (Mancuso et al. 2001). In those patients achieving complete remission after chemotherapy, the number of CECs was similar to healthy controls (Mancuso et al. 2001). Accordingly, an increased number of CECs has been recognized in younger patients and those with aggressive NHL and the levels of CECs decreased following complete response to treatment (Igreja et al. 2007).

3.7 Genetically Modified Lymphoma Endothelial Cells

Chromosomal abnormalities involving all chromosomes may occur in lymphomas and characterization of genetic abnormalities, while not an absolute requirement, can be essential to the diagnosis of many lymphomas (Seto 2010).

The presence of lymphoma-specific chromosomal translocations in endothelial cells in B cell lymphomas has been demonstrated, suggesting that microvascular endothelial cells in B cell lymphomas are in part tumor related (Streubel et al.

2004). Moreover, 15–85 % of microvascular endothelial cells harboured lymphoma-specific genetic alterations consisting not only of B cell specific translocation of immunoglobulin heavy chain (IGH), but also secondary genetic alterations in FL (Streubel et al. 2004). As suggested by Streubel et al. (2004) four mechanisms may be involved: lymphoma cells and endothelial cells may derived from a multipotent hemangioblastic precursor cell targeted by neoplastic transformation which can differentiate in tumor cells or endothelial cells sharing the same genetic abnormalities; the endothelial cells carrying the genetic alterations of the lymphoma may arise from a cell that was already committed to the lymphoid lineage; fusion of lymphoma cells and endothelial cells with formation of hybrid vessels or gene transfer by means of the uptake of apoptotic bodies from tumor cells by neighboring cells may be alternative mechanisms.

The expression of a transcript called T cell Ig and mucin-containing molecule 3 (Tim-3) has been identified in microvessels of DLBCL, but not in reactive lymph nodes, suggesting that the lymphoma endothelium may act as a functional barrier facilitating the establishment of lymphoma immune tolerance (Huang et al. 2010).

Chapter 4
Angiogenesis in Leukemia

4.1 General Features of Leukemias

B cell acute lymphoblastic leukemia (ALL) originates from the clonal expansion of early B cells progenitors, while chronic lymphocytic leukemia (CLL) is a chronic lymphoid malignancy characterized by accumulation of $CD5^+$ monoclonal B lymphocytes in peripheral lymphoid organs, bone marrow, and peripheral blood (Pui et al. 2004).

Reciprocal positive and negative interactions occur in the bone marrow between leukemic blasts and bone marrow stromal cells, that are mediated by an array of angiogenic growth factors and soluble mediators (Litwin et al. 2002). These interactions modulate the angiogenic response, the proliferation of leukemic cells, the development of disease and the acquisition of drug resistance (Burger et al. 2009).

Angiogenesis, favoured by the hypoxic bone marrow environment, increases the oxygen tension in the bone marrow and stimulates leukemic cells tumor growth. Low oxygen tension modulates the expression of HIF-1, which also regulates angiogenesis-related genes, including the protein expression of angioregulatory cytokines. In this context, hypoxia might thereby influence leukemogenesis and chemosensitivity in childhood ALL and acute myeloid leukemia (AML) (Wellman et al. 2004; Hatfield et al. 2010a).

4.2 Angiogenesis in Acute Lymphocytic Leukemia

In 1997, Perez-Atayde et al. in a study of 51 children with ALL, demonstrated a high bone MVD, and an altered and chaotic spatial organization of the tumor vasculature. Moreover, tumor cells were arranged around new-formed microvessels and urinary levels of FGF-2 decreases to normal levels after a complete response to therapy (Perez-Atayde et al. 1997). Pulè et al. (2002) confirmed these findings in a study on 41 patients treated with standardized protocols with a mean follow-up of 4 years. Microvascular density in most patients was significantly higher at

D. Ribatti, *Angiogenesis and Anti-Angiogenesis in Hematological Malignancies*,
DOI 10.1007/978-94-017-8035-3_4, © Springer Science+Business Media Dordrecht 2014

presentation and decreases in remission, while MVD was comparable in children with poor prognosis and in those who relapsed. Moreover there was no associations with age, sex, cytogenetic abnormality or disease phenotype (Pulè et al. 2002). Increased bone marrow MVD and increased plasma levels of FGF-2 and VEGF levels have been correlated with shorter overall and disease-free survival (Aguayo et al. 2000). Bone marrow plasma from most all ALL patients contained elevated levels of FGF-2 compared to the bone marrow plasma of normal donors (Veiga et al. 2006). In contrast, the levels of VEGF did not significantly differ between leukemic and normal bone marrow plasma. This discrepancy between serum levels of FGF-2 and VEGF could depend by the fact that there is a lot of VEGF in platelets and ALL patients at diagnosis are thrombocytopenic. Serum level of an endogenous anti-angiogenic factor, namely endostatin, was significantly higher in childhood ALL than control at diagnosis and in remission (Schneider et al. 2007).

As concerns the association with prognostic factors and the outcome of patients, significantly higher VEGF levels in patients at relapse, and lower levels at diagnosis in association with a longer overall survival have been found (Koomagi et al. 2001). Yetgin et al. (2001) reported higher VEGF levels in complete remission than at diagnosis, suggesting that VEGF could increase when normal hematopoiesis is restored. Schneider et al. (2001) demonstrated that the levels of FGF-2 were markedly elevated in ALL children versus controls, but lower for patients with poor prognostic features. On the contrary, VEGF levels were significantly increased in relapsing patients (Schneider et al. 2003). Norén-Nyström et al. (2009) found correlations between higher MVD and T-ALL phenotype, high- hyper diploid leukemia (a criteria of good prognosis), and with blood cells count in childhood B-ALL, and showed that B-ALL patients had a worse outcome when having high MVD. These findings indicate that the prognostic value of angiogenesis may differ according to the subtypes of childhood ALL and is coupled to both the ALL genotype and phenotype.

Overall, these data confirm the important and crucial role of angiogenesis in bone marrow of patients with ALL, the involvement of VEGF and FGF-2 as principal angiogenic cytokines, and it prognostic value.

4.3 Angiogenesis in Chronic Lymphocytic Leukemia

Chronic lymphocytic leukemia is a post-germinal center neoplasms characterized by clonal proliferation and accumulation of mature appearing lymphocytes in the blood, bone marrow, lymph nodes and spleen (Table 4.1). Chronic lymphocytic leukemia is the most common leukemia in Western countries, and it account for approximately one-third of all leukemias in the United States.

Approximately, 50% of freshly isolated CLL cells expressed mRNA and protein for VEGF (Dias et al. 2000). Moreover, VEGF released by CLL cells stimulated endothelial cell proliferation *in vitro* in the Matrigel assay, conditioned media from cultured CLL cells stimulated angiogenesis in the CAM assay, and this was inhibited by a blocking anti-VEGF antibody (Kini et al. 2000). CLL cell-derived

Table 4.1 Classification of chronic lymphocytic leukemia accordingly Rai and Binet staging systems

Rai stage	
O	Low
I	Intermediate
II	Intermediate
III	High
IV	High
Binet stage	
A	Low
B	Intermediate
C	High

microvesicles induced VEGF production in bone marrow stromal cells through the activation of AKT/mTOR/p7056/HIF-1α axis, and VEGF production could be inhibited by rapamycin inhibitor (Ghost et al. 2010). Treatment with exogenous VEGF caused a reduction in both spontaneous and drug induced apoptosis in CLL cells (Lee et al. 2005). Leukemia cells also express VEGFR-2 and *in vitro* using RT-PCR and Northern blot analysis, VEGFR-2 was detected, and it was shown to be functional *in vivo* (Bellamy et al. 1999). VEGFRs of CLL cells are constitutively phosphorylated and this phosphorylation can be blocked by addition of VEGF antibodies (Lee et al. 2005). Moreover, VEGF signaling via VEGFR-1 and VEGFR-2 in CLL cells confers resistance to apoptosis and promotes survival. Blocking of VEGF and VEGFRs by using neutralizing antibodies or tyrosine kinase inhibitors, resulted in decreased levels of p-STAT-3 and apoptosis of CLL cells (Lee et al. 2005). In co-cultures of CLL B cells and marrow stromal cells a significant increase in the secretion of FGF-2 or VEGF and a decrease of anti-angiogenic TSP-1 was demonstrated after 24 h (Panayiotidis et al. 1996; Edelman et al. 2008).

Angiogenesis is involved in the pathogenesis of B-cell CLL (Aguayo et al. 2000; Kini et al. 2000; Molica et al. 1999, 2000, 2002; Chen et al. 2000). Enhanced vessel density was observed in CLL infiltrated bone marrow and lymph nodes compared with controls (Kini et al. 2000; Chen et al. 2000). Intracellular FGF-2 levels in lymphocytes of B-CLL patients correlated with stage of disease (Menzel et al. 1996; König et al. 1997), and a positive correlation was identified between serum FGF-2 levels and Bcl-2 expression (König et al. 1997; Bairey et al. 2000), resulting in increased leukemia cell survival and increased cellular FGF-2 in high-risk CLL was associated with fludarabine resistance (Menzel et al. 1996).

Baban et al. (1996) first reported the elevated expression of VEGF in CLL patients, measured in both serum and leukemic cells (Kini et al. 2000; Chen et al. 2000) and significant amounts were produced under hypoxic conditions (Chen et al. 2000; Kay et al. 2002). Isoforms predominantly expressed are VEGF121 and VEGF165, and VEGF is produced by both circulating and tissue-phase CLL cells, providing direct evidence of its angiogenic effects (Chen et al. 2000). Bone marrow microvascular density and urinary VEGF levels were not different in CLL patients and controls and did not correlate with the clinical stage, despite increased serum levels of angiogenic factors (Aguayo et al. 2000; Chen et al. 2000). Molica et al. (2002)

showed that MVD was significantly higher in bone marrow biopsies of newly diagnosed patients with B-CLL than those of controls, but it was not correlated other predictors representative of tumor mass or disease progression. Low cell and high serum VEGF levels correlated with a poor clinical outcome (Aguayo et al. 2000; Molica et al. 1999; Smolej et al. 2007). Intracellular VEGF levels measured by Western blot analysis and solid-phase radioimmunoassay were significantly higher in B-CLL cells than in normal peripheral blood mononuclear cells (Aguayo et al. 2000). In B cell CLL, serum levels of VEGF reflected an aggressive profile and correlated positively with prognostic markers, including ZAP-70, CD38, and mutational status of immunoglobulin variable gene segments (IgVhH) (Molica et al. 2007b). A clinical study conducted on 83 untreated patients demonstrated that serum concentrations of VEGF and VEGFR-2 were significantly higher in advanced stages disease (Gora-Tybor et al. 2005).

CLL cells express PDGF as well as PDGF receptor (PDGFR), and PDGF levels were found to be significantly elevated in CLL plasma compared with normal plasma and were strongly correlated with VEGF levels in CLL patients with high risk features (Ding et al. 2010). Moreover, Ding et al. (2010) demonstrated that both PDGF and VEGF were elevated in plasma of CLL patients with a positive association for high-risk factors and more advanced stage. Moreover, PDGF induced VEGF production in mesenchymal stromal cells.

Maffei et al. (2010a) demonstrated that in CLL increased expression of Ang-2 was positively correlated with enhanced bone marrow angiogenesis and that high Ang-2 level was an independent prognostic factor for shorter time to first treatment. Moreover, Maffei et al. (2010b) found a significant association between high level of Ang-2 and advanced clinical stage, high β2-microglobulin, un-mutated status of the IgVH and cytogenetic risk factors.

4.4 Angiogenesis in Acute and Chronic Myeloid Leukemia and Myelodysplastic Syndrome

In AML adult patients, high levels of cellular VEGF were significantly correlated with shorter survival in patients with high white blood cells at diagnosis and with high number of circulating blasts, whereas no association was found with the plasma levels of FGF-2 (Aguayo et al. 1999). Fiedler et al. (1997) found that a large proportion of AML patients expressed VEGF and VEGFRs. VEGFR-1 expression levels were found to be similar between AML bone marrow samples and normal bone marrow, while VEGFR-2 was more commonly expressed on AML myeloblasts (Padro et al. 2002) and blocking VEGFR-2 antibodies resulted in decreased leukemic cell growth (Dias et al. 2000). The overall VEGFR-3 protein expression levels were significantly higher in AML patient bone marrow samples, as compared to controls (Liersch et al. 2008). Enhanced VEGF plasma levels were associated with lower complete remission rates and a reduced survival in AML patients (Aguayo et al. 2002; De Bont et al. 2002; Wegiel et al. 2009).

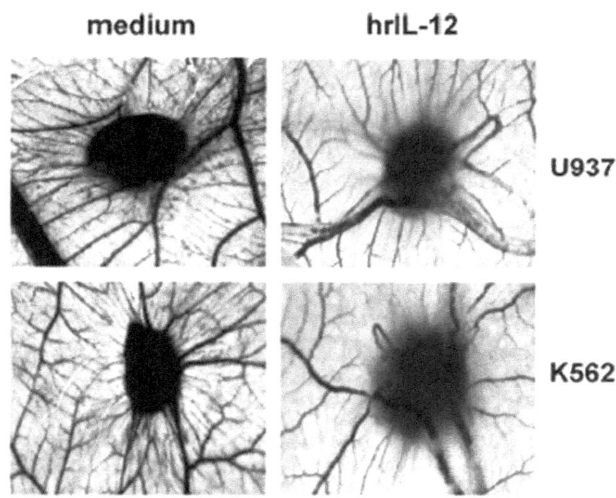

Fig. 4.1 Angiogenic activity of supernatants from two acute myeloid leukemia cell lines, U937 (*upper panels*) or K562 (*lower panels*), cultured in the absence (*left panels*) or presence (*right panels*) of rhIL-12, as assessed by the *in vivo* CAM assay. (Reproduced from Ferretti et al. 2010)

Hussong et al. (2000) found significantly increased MVD in the bone marrow of AML patients and higher MVD predicted for poor prognosis (Kuzu et al. 2004; Rabitsch et al. 2004).

Upon transwell or direct co-culture with endothelial cells, AML blasts proliferate to a higher degree and are less susceptible to chemotherapeutics agents (Liesveld et al. 2005; Hatfield et al. 2009). *In vitro*, VEGF stimulation increases the AML blast proliferation in synergy with stem cell factor (SCF) (Foss et al. 2001) and promotes the survival of AML cells (Dias et al. 2000). The pro-survival signaling mechanism protect AML cells from chemotherapy-induced apoptosis through up-regulation of anti-apoptotic proteins (Dias et al. 2000). In pediatric AML various isoforms of VEGF, including VEGF121, VEGF165, and VEGF189, are expressed in AML cells (Kruizinga et al. 2011).

Primary human AML cells induced a strong angiogenic response *in vivo* in the CAM assay (Fig. 4.1) (Ferretti et al. 2010), expressed Ang-1, Ang-2, and their receptor, Tie-2 (Wakabayashi et al. 2004; Hatfiled et al. 2008), and blocking Angs interactions with Tie-2 decreased AML cell proliferation (Reikvam et al. 2010a). Hatfiled et al. (2012) used a co-culture assay of endothelial and vascular smooth muscle cells to investigate the effects of AML-conditioned medium and demonstrated that it stimulated endothelial cell organization into capillary-like networks. Other cytokines are constitutively released by human AML cells including chemokines [CCL2-4/CXCL 1-8; CCL5/CXCL9-11; CCL13-17-22-24/CXCL5 (Bruserud et al. 2007)], and HGF. This system constitute a functional interacting network which exert direct and indirect effects on the leukemic cells, regulating the crosstalk between leukemic and stromal cells (Reikvam et al. 2013). Nuclear factor kB (NFkB) is a transcription factor known to affect the expression of several chemokines and it has been demonstrated that bortezomib, a proteasome inhibitor that inhibits NFkB activation decreases CCL2 and CCL4 levels (Bruserud et al. 2007). In this context, the chemokine system may be considered as a possible therapeutic target and chemokine inhibitors used as a class of drugs that may become important for inhibition of angiogenesis in leukemias (Melve et al. 2011).

In chronic myeloid leukemia (CML), the number of VEGF positive bone marrow cells is significantly higher than controls and correlate with bone marrow vascularity (Lundberg et al. 2000), and VEGFR-2 overexpression is an independent prognostic indicator for shortened survival in patients with CML (Verstovsek et al. 2002). Tie-1 is an independent predictor of survival in early chronic phase CML (Versovsek et al. 2002b), and the bcr-abl construct has been identified in the endothelial cells of bone marrow microvessels in patients with CML (Gunsilius et al. 2000), suggesting that these microvessels are involved in the pathogenesis of the disease. In fact, *in vitro* studies have demonstrated increased secretion of VEGF by tumor cells transfected with the bcr-abl construct (Mayerhofer et al. 2002) and the secretion can be inhibited by the bcr-abl specific tyrosine kinase inhibitor STI571 (imatinib mesylate) (Ebos et al. 2002).

In myelodysplastic syndrome (MDS), MVD has been investigated and VEGF and VEGFR-2 have been detected in myeloblasts and their expression was higher and correlated with MVD in high risk as compared to low risk patients (Bellamy et al. 2001b; Korkolopoulou et al. 2001).

4.5 Role of Matrix Metalloproteinases and Mast Cells

The levels of tissue inhibitor of metalloproteinase-1 and -2 (TIMP-1 and TIMP-2) and of MMP-2 were significantly higher, while MMP-9 levels were significantly lower in the bone marrow of ALL patients than in controls (Lin et al. 2002).

CLL cells could attack to and bound both pro-MMP-9 and active MMP-9 through a docking complex constituted by $\alpha4\beta1$ and CD44v and this interaction was mediated by the MMP-9 hemopexin domain (PEX-9) (Redondo-Munoz et al. 2008). High plasma MMP-9 levels and adult T cell leukemia (ATL) cell invasion, as well as plasma VEGF levels are closely related (Hayashibara et al. 2002). Production of MMP-2 and MMP-9 by CLL B-cells may be suppressed by interferons (Kay et al. 2002; Bauvois et al. 2002). Immunohistochemical evidence demonstrated that MMP-9 was associated with areas of tissue remodeling and invasion (Kamiguti et al. 2004) and elevated intracellular levels of MMP-9 correlated with advanced CLL stage and poor survival. No correlation was found between MMP-9 and bone marrow angiogenesis (Molica et al. 2003a).

Primary human AML cells showed constitutively release of both MMPs and TIMPs, which are important for leukemogenesis (Reikvam et al. 2010b). Moreover, in AML where both chemokines and MMPs can be constitutively secreted, MMPs can activate angiogenic chemokines through proteolytic cleavage (Hatfiled et al. 2010b).

A correlation was found between the extent of angiogenesis and the number of tryptase-reactive mast cells in B-cell CLL on tissue sections selected from Binet

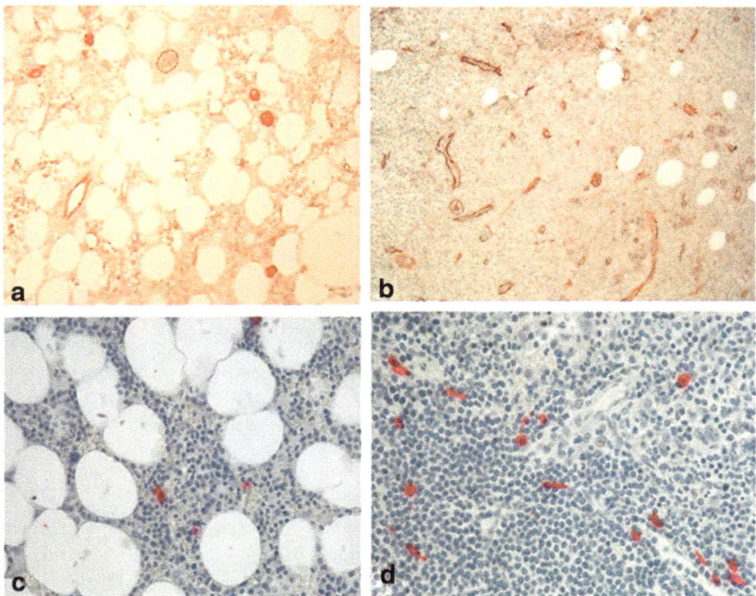

Fig. 4.2 Sections of bone marrow biopsies stained with Factor VIII-RA for microvessels (**a**, **b**) and with tryptase for mast cells (**c**, **d**) from a control subjects (**a**, **c**) and from a patient with B-cell chronic lymphocytic leukemia (**b**, **d**). Note in **a** and **b** strongly stained megakaryocytes as internal positive controls. (Reproduced from Ribatti et al. 2003b)

stage A patients (Ribatti et al. 2003b). Mast cells were generally scattered throughout the neoplastic tissue and rested near or around the blood capillaries within the interstitial stroma (Fig. 4.2) (Ribatti et al. 2003b). Tryptase-positive mast cells number did not correlate with Rai substage, peripheral blood lymphocytosis and β2-microglobulin, and disease progression (Molica et al. 2003b). The consistent decrease of bone marrow angiogenesis after sequential fludarabine-induction and alentuzumab-consolidation therapy in advanced CLL paralleled to the reduction of mast cells in bone marrow (Molica et al. 2007a).

Chapter 5
Anti-angiogenesis

5.1 Introduction

In 1971, J Folkman published, in the *New England Journal of Medicine*, a hypothesis that tumor growth is angiogenesis dependent. The hypothesis predicted that tumors would be unable to grow beyond a microscopic size of 1–2 mm^3 without continuous recruitment of new capillary blood vessels. Folkman introduced the concept that tumors probably secreted diffusible molecules that could stimulate the growth of new blood vessels toward the tumor and that the resulting tumor neovascularization could conceivably be prevented or interrupted by drugs called angiogenesis inhibitors.

Beginning in the 1980s, the biopharmaceutical industry began exploiting the field of anti- angiogenesis for creating new therapeutic compounds for modulating new blood vessel growth in angiogenesis-dependent diseases. In the last 35 years, it has been estimated that >200 companies have worked and are still working in the area of angiogenesis and several of the compounds that modulate angiogenesis are currently being evaluated in clinical trials. A list of approved anti-angiogenic drugs approved for clinical use is available in the Table 5.1.

5.2 Endostatin

Endogenous inhibitors of angiogenesis are defined as proteins or fragments of proteins tha are formed in the blood and can inhibit angiogenesis. Various endogenous inhibitors of angiogenesis may be found in the bloodstream and a circulating form of human endostatin has been identified (Standker et al. 1997). Endostatin is an endogenous inhibitor of angiogenesis, which inhibits endothelial cell proliferation and migration, induces apoptosis and causes a G1 arrest of endothelial cells. Moreover, endostatin inhibits MMP-2 activity, blocks the binding of VEGF to VEGFR-2, and stabilizes cell-cell and cell-matrix adhesions, preventing the breakage of these junctions required during angiogenesis (Folkman 2006).

Angiostatin and endostatin inhibit bone marrow angiogenesis and leukemia in experimental animals (Scappaticci et al. 2001). In a subgroup of patients with large

D. Ribatti, *Angiogenesis and Anti-Angiogenesis in Hematological Malignancies*,
DOI 10.1007/978-94-017-8035-3_5, © Springer Science+Business Media Dordrecht 2014

Table 5.1 List of antiangiogenic drugs approved for clinical use

Drug	Target	Company	Indication
Avastin (Bevacizumab)	VEGF	Genentech	mCRC, NSCLC, Advanced breast cancer
Erbitux (Cetuximab)	EGFR	Imclone	mCRC & Head and Neck cancer
Vectibix (Panitumumab)	EGFR	Amgen	mCRC
Lucentis (Ranibizumab)	VEGF	Genentech	Wet Age-related macular regeneration
Macugen (Pegaptanib)	VEGF	OSI Pharmaceuticals	Wet Age-related macular regeneration
Endostar (Endostatin)	Angiogenesis inhibitor	Shangdong Simcere Medgen	Lung cancer
Sorafenib (Nexavar)	VEGFR, PDGFR & Raf	Bayer AG/Onyx	Advanced RCC
Sunitinib (Sutent)	VEGFR, PDGFR & c-kit	Pfizer	Advanced RCC & GIST
Thalomid (Thalidomide)	Angiogenesis inhibitor	Celgene Corporation	Multiple Myeloma
Sorafenib (Nexavar)	VEGFR, PDGFR & c-kit	Bayer AG/Onyx	Advanced RCC
Sunitinib (Sutent)	PDGFR & VEGFR	Pfizer	Advanced RCC & GIST
Dasatinib (Sprycel)	Bcr-Abl & Src	Bristol-Myers Squibb	Gleevec-resistant CML or Ph+ ALL
Lapatinib (Tykerb)	EGFR & Her2/neu	GlaxoSmithKline	Advanced metastatic Her2+ breast cancer
Velcade (Bortezomib)	Proteosome inhibitor	Millenium Pharmaceuticals	Multiple myeloma
Tarceva (Erlotinib)	EGFR	Genentech/OSI	Lung cancer

cell and immunoblastic lymphoma, patients with high serum endostatin levels had a significantly better survival as compared with those with lower levels (Bono et al. 2003). Bertolini et al. (2000) showed that endostatin-induced tumor stabilization after chemotherapy or anti-CD20 immunotherapy in a mouse model of human high-grade NHL. Moreover, continuous infusion of endostatin inhibits tumor growth and the mobilization and differentiation of EPCs in mice bearing an angiogenic human lymphoma (Capillo et al. 2003).

Treatment of lymphoma-bearing mice with endostatin caused an increase in the frequency of apoptotic cells in the endothelial cell compartment and most of CECs were apoptotic or dead, while cyclophosphamide had no such effect. This difference probably occurred because most of the circulating apoptotic cells were hematopoietic and not endothelial in nature (Monestiroli et al. 2001). Endostatin administration in advanced stages of tumor growth leaded to tumor regression even in cyclophosphamide and rituximab resistant cases (Bertolini et al. 2000). This effect was induced by inhibition of proliferation and stimulation of apoptosis in endothelial cells.

5.3 Thalidomide

The exact reason why thalidomide causes limb defects remained unknown until Folkman's group identified its anti-angiogenic effects (Ribatti and Vacca 2005). Given that development of foetal limbs is heavily dependent on the formation of new blood vessels, this is at least in part the reason why such defects were observed. D'Amato et al. (1994) used a rabbit cornea micropocket assay in which angiogenesis was induced by FGF-2 in order to demonstrate that thalidomide inhibits angiogenesis mainly by directly preventing new vessel formation as opposed to suppression of infiltrating host inflammatory cells. Histological sections of the pretreated neovascularized corneas were virtually free of inflammatory cells. Later, the same group reported that thalidomide inhibited VEGF in a murine model of corneal vascularization (Kenyon et al. 1997) and others have shown that it inhibits microvessel formation in a rat aorta ring assay (Raver et al. 1998) Moreover, Verheul et al. (1999) found that thalidomide inhibits the growth of V2 carcinoma in rabbits by an anti-angiogenic mechanism. The precise mechanism of action of thalidomide has been difficult to elucidate because of its relative lack of activity *in vitro* compared to effects *in vivo*, and the fact that in solutions at physiologic pH, it undergoes spontaneous nonenzymatic hydrolysis into numerous metabolities. The immunomodulatory effects of thalidomide depend on the drug's inhibition of TNF-α (Hideshima et al. 2001b) and suppression of NF-kB activation in response to inflammatory agents (Keifer et al. 2001; Majunbar et al. 2002). By acting as a costimulator, thalidomide increases the response of T cells to T cell receptor-mediated stimulation, yielding increased proliferation and greater production of IL-2, IL-10 and IFN-γ (Shannon et al. 1995; Partida-Sanchez et al. 1998). Thalidomide increases the number of natural killer cells, while these cells exhibit augmented cytotoxic activity against tumor cells (Davies et al. 2001). In addition, it downregulates IL-5, IL-6, IL-8 and IL-12 production (Moller et al. 1997; Moreira et al. 1997; Rowland et al. 1998). A downregulation of adhesion molecules and a decrease in cell–cell contact between human T leukaemic cells and HUVECs cells has been demonstrated after exposure to thalidomide (Settles et al. 2001). Adhesion molecules are involved by facilitating the adherence of tumor cells to the endothelium and are required for the development of metastases. Thalidomide also induces tumor cell apoptosis, as evidenced by increased sub-G1 cells for the induction of p21 and growth related arrest (Hideshima et al. 2000).

5.3.1 Thalidomide in the Treatment of Multiple Myeloma and Waldenstrom's Macroglobulinemia

Since thalidomide possessed anti-angiogenic properties, Singhal et al. (1999) used it on a compassionate basis to treat 84 patients with relapsed and refractory MM. Thalidomide was thus the first new drug to display significant single-agent activity against relapsed and relapsed refractory MM in more than 30 years. It was administered for a median of 80 days at an initial dose of 200 mg taken at night, augmented by 200 mg every 2 weeks to a maximum of 800 mg. Response was defined as reduction in serum or urine levels of paraprotein and was typically apparent within 2 months. A 90 % reduction was documented in 8 patients (with two achieving complete response), as reflected by disappearance of paraprotein in both serum and urine. Decreases in plasma cell infiltration of bone marrow and increased haemoglobin values were observed in 78 % of the responders, although no difference in the bone marrow MVD was found between responsive and resistant patients. However, only a subset of patients had bone marrow serially assessed for angiogenesis, limiting the strength of this observation to truly rule out any anti-angiogenic effect against new vessel formation. While inhibition of new vessel formation remains an attractive potential mechanism of action, thalidomide has a number of other properties that may explain its activity in MM. These include disruption of adhesion molecule expression and as a result, alteration of the interaction of myeloma cells with the bone marrow microenvironment (Geitz et al. 1996). This decreases the secretion of cytokines that augment growth and survival (Chauhan et al. 1996; Hallek et al. 1998). and drug resistance (Damiano et al. 1999). Thalidomide also induces dose-dependent modulation of TNF-α production by stimulated monocytes, macrophages and neutrophils (Sampaio et al. 1991) with no effect on total protein synthesis or the production of other cytokines. Finally, thalidomide has been shown to enhance cell-mediated immunity by direct costimulation of T cells (Mc Hugh et al. 1995). These effects are also accompanied by means of the T cell receptor complex, thereby increasing IL-2-mediated cell proliferation and IFN-γ production. IL-6 is a major growth survival and drug-resistant factor in MM (Klein et al. 1995) and its suppression by thalidomide inhibits plasma cell growth. Other studies have shown that thalidomide is active in relapsed MM with response rates ranging from 25 to 35 % (Rajkumar et al. 2000a; Barlogie et al. 2001; Dimopoulos et al. 2003; Kumar et al. 2003b) and a median response duration of approximately 12 months (Barlogie et al. 2001; Dimopoulos et al. 2003; Kumar et al. 2003b). Thalidomide has also shown to be useful in early stage of myeloma and is being assessed for the treatment of both newly diagnosed symptomatic patients and those with smouldering disease (Rajkumar et al. 2001; 2002a, b; Weber et al. 2003). Mohty et al. (2005) demonstrated that single agent thalidomide is potentially effective in MM patients who progress after allogenic stem cell transplant, confirming a prior study by Richardson et al. (2004b) in which both patients relapsing after autologous stem cell transplant and allogenic stem cell transplant responded to dose escalation. In this latter study, dose dependent as well as cumulative toxicity was observed with the authors concluding that doses should be individualized according to response and tolerance with a median dose of 200 mg/day administered in this study.

A 25 % partial response was obtained in prospective phase II study of thalidomide in patients with symptomatic Waldenstrom's macroglobulinaemia (Dimopuolos et al. 2001b) and a 30 % response was achieved when it was combined with anti-CD20 monoclonal antibody (Desikan et al. 2002). BLT-D, a nonmyelosuppressive combination composed of clarithromycin, low-dose thalidomide and dexamethasone, has also been used (Coleman et al. 2003).

5.3.2 Thalidomide in the Treatment of Leukemia, Lymphoma and Myelodysplastic Syndrome

In CLL, Chanan-Khan et al. (2005) in a phase I trial by using 6 months of continuous daily thalidomide with standard monthly doses of fludarabine on 13 previously untreated patients, obtained a 100% overall response rate with 55% complete remission rate.

In AML patients, thalidomide was tested alone or in combination with other compounds (Thomas et al. 2003; Steins et al. 2002; Barr et al. 2007). Microvascular density significantly decreased during treatment with thalidomide (Rajkumar et al. 2008) and serious adverse effects were observed when thalidomide was administered in combination with fludarabine, carboplatin, and topotecan (Chanan-Khan et al. 2005).

Thalidomide's metabolites have cytotoxic effects on leukemic cells through induction of their morphological differentiation *in vitro* (Hatfill et al. 1991). The efficacy and tolerability of thalidomide in refractory and relapsed acute and CML have been assessed. A phase I–II trial conducted by Steins et al. (2002) during which 200–400 mg thalidomide daily were administered for 7 weeks (on average) produced a 39% partial response characterized by 450% reduction of blasts and significant lowering of bone marrow MVD. Serum levels of FGF-2 and VEGF levels were also reduced.

Thalidomide as single agent demonstrated a low overall response rate in patients with relapsed/refractory indolent NHL (Smith et al. 2008) and in heavily pre-treated patients with recurrent lymphoma (Pro et al. 2004). In combination with fludarabine, thalidomide was associated with significant therapeutic efficacy in CLL (Chanan-Khan et al. 2005).

Wilson et al. (2002) evaluated thalidomide (200 mg daily with a 100 mg daily increment to a maximum of 800 mg daily for 2 months) in a patient with mantle-cell NHL resistant to conventional chemotherapy. The complete blood count (CBC) was improved and there was a 50% reduction of splenomegaly and abdominal lymphoadenopathy. Bone marrow osteomedullary biopsies revealed a distinct decrease in tumor infiltrate.

Raza et al. (2001) obtained a 41% partial remission with thalidomide plus topotecan, pentoxyphylline and/or etanercept (a subcutaneously administered biological response modifier that binds and inactivates TNF-α). In a subsequent study, thalidomide was given at 100–400 mg daily for 12 weeks (Raza et al. 2001). Approximately, 50% of patients achieved a partial remission with an increase in haemoglobin, platelets, and absolute neutrophil count, improvement of marrow dysplasia and no need for blood transfusions. A decrease in serum TNF-α and bone marrow MVD provided further evidence of the drug's anti-angiogenic activity. Strupp et al. (2002a) studied 34 patients who received 100 mg thalidomide daily with a weekly increase of 100 mg daily to 400 mg daily (the maximum-tolerated dose): 66% responded, 20% progressed and 14% remained stable.

5.4 Side Effects

Side effects associated with thalidomide include sedation, fatigue, constipation, tremor, dizziness, bradycardia and skin rashes. These effects are generally mild, but may affect a patient's quality of life and lead to the discontinuation of treatment, especially when high doses are used. Peripheral neuropathy also limits long administration. Deep vein thrombosis (DVT) is usually observed when thalidomide is administered to previously untreated patients in combination with dexamethasone, doxorubicin or melphalan. Its incidence rises from only 1 to 3% for thalidomide alone, to between 10 and 15% when combined with dexamethasone, and up to approximately 25% when thalidomide is combined with other cytotoxic agents, particularly doxorubicin. Side effects are typically reversible and disappear after decreasing dose and/or drug suspension. In the Singhal study, median duration of treatment with thalidomide was 80 days and most patients received daily doses of 400 mg, with fewer retaining 600 and 800 mg daily (Shinghal et al. 1999). Subsequent studies have shown doses of 100–200 mg/day being better tolerated (Richardson et al. 2002a). minimize complications including weakness, fatigue and somnolence, thalidomide is typically administered as a onetime dose in the evening, with a gradual increase of the dose to response and tolerability as the risk of peripheral neuropathy increases both with prolonged exposure and higher dose. Given the risk of accumulative toxicity and neuropathy, particularly with the advent of other agents that are both effective but nonetheless have the possibility of overlapping toxicity with thalidomide, prolonging therapy in the context of established neuropathy, for example, should be avoided (Richardson et al. 2002).

5.5 Thalidomide Analogues

Two classes of thalidomide analogues have been synthesized, namely phosphodiesterase 4 inhibitors that inhibit TNF-α but do not enhance T cell activation (selected cytokine inhibitory drugs: SelCiDs) and analogues that stimulate T cell proliferation as well as IL-2 and IFN-γ production (immunomodulatory drugs: IMiDs) (Corral et al. 1999). The activity of thalidomide and that of its metabolities on HUVECs proliferation overlap *in vitro*, whereas the metabolities are four to five times more active inhibitors of angiogenesis in the CAM assay (Marks et al. 2002). Two IMiDs, CC-5013 (Revlimid, lenalinomide) and CC-4047 (Actimid), display antimyeloma activity (Richardson et al. 2004a; Schey et al. 2004). By modifying thalidomide structure through the addition of an amino group at the fourth position of the phthaloyl ring (to generate CC-4047) and with the further removal of a carbonyl on the ring to form CC-5013, such analogues are up to 50,000 times more potent at inhibiting TNF-α than the thalidomide parent compound *in vitro* and are markedly more stable (Richardson et al. 2002b). CC-5013 is clinically active even in thalidomide-resistant patients, is not teratogenic in *in vivo* preclinical models and its adverse-event profile is free from thalidomide's tendency to produce peripheral neuropathy, somnolence and constipation (Richardson et al. 2002b). Its proposed mechanisms of action also include direct antiproliferative and proapoptotic effects on MM cells through inhibition of NF-kB transcriptional activity and activation of death receptor/caspase-8-mediated death signaling (Mitsiades et al. 2002b) modulation of MM–bone marrow stromal cell adhesive interactions, abrogation of secretion of prosurvival cytokines (Hideshima et al. 2000), increase of natural killer cell number and cytotoxic activity against MM cells (Davies et al. 2001). It is thus a promising agent for the treatment of MM and trials conducted for approval by the Food and Drug Administration (FDA) are under way. The first phase I study of CC-5013 was carried out in 2001, when a dose range of 5–50 mg daily was tested in 25 patients with relapsed and refractory MM (Richardson et al. 2002b). No significant constipation or neuropathy was seen and encouraging responses were seen in 17 (71 %) patients, including 11 patients (46 %) who had received prior thalidomide. Additional clinical phase II trials have confirmed these data, achieving impressive responses with a favourable side-effect profile, and two clinical phase III trials comparing CC-5013 combined with high-dose dexamethasone to high-dose dexamethasone alone have been completed (Podar and Anderson 2005). Preliminary analysis has confirmed markedly superior outcome for the combination both in terms of time to progression and response rate, but the side effect profile was noteworthy for a high rate of DVT and pulmonary embolus at 14 %, a finding which was a surprise given the absence of thromboembolic complications seen with CC-5013 alone. The effects of combining CC-5013 with rapamycin, a specific mTOR inhibitor, have been studied preclinically, based on rational targeting of the differential signaling cascade-mediating MM cell growth and survival (Raje et al. 2004). Results have shown synergy between these two orally bioavailable agents and suggest that low-dose combinations may be effective. It should be noted that this *in vitro* work has not yet been tested *in vivo*. Clinically, studies will need to be approached with caution due to the fact that both of these agents are toxic for endothelial cells and thus vascular injury may be considerable. In a report, six out nine patients with myelodysplastic syndrome treated with CC-5013 at the dose of 25 mg daily achieved haematological benefits, including transfusion independence, reduction of bone marrow blasts and cytologic dysplasia and improvement in marrow multipotent and erythroid progenitor formation (List et al. 2005). Schey et al. (2004) reported data from a phase I study addressing the tolerability, efficacy and immunologic effects of CC-4047 in the treatment of patients relapsed or relapsed refractory MM, and results suggest that this agent appears very active but is associated with significant toxicity, including DVT, even when used as a single agent.

Lenalidomide inhibits MM plasma cells induced angiogenesis *in vivo* in the CAM assay and endothelial cells induced angiogenesis *in vitro* in the Matrigel assay (Fig. 5.1) (De Luisi et al. 2011). Moreover, lenalidomide inhibits MM endothelial cell migration, not affect MM endothelial cell viability and adhesion and downregulates key genes and proteins related to MM angiogenesis (Figs. 5.2 and 5.3) (De Luisi et al. 2011).

Lenalidomide has been used as monotherapy in the treatment of both aggressive (DLBCL and transformed) and indolent relapsed/refractory NHL, MCL, and angio-immunoblastic T cell lymphoma (Strupp et al. 2002b; Dogan et al. 2005; Wiernik et al. 2008; Habermann et al. 2009; Witzig et al. 2009; Cuzcman et al. 2011). Lenalidomide has also been studied in relapsed/refractory CLL inducing a complete and partial remission and have considerable activity in both heavily pre-treated CLL patients and patients with unfavorable prognostic factors (Chanan-Khan et al. 2006; Ferrajoli

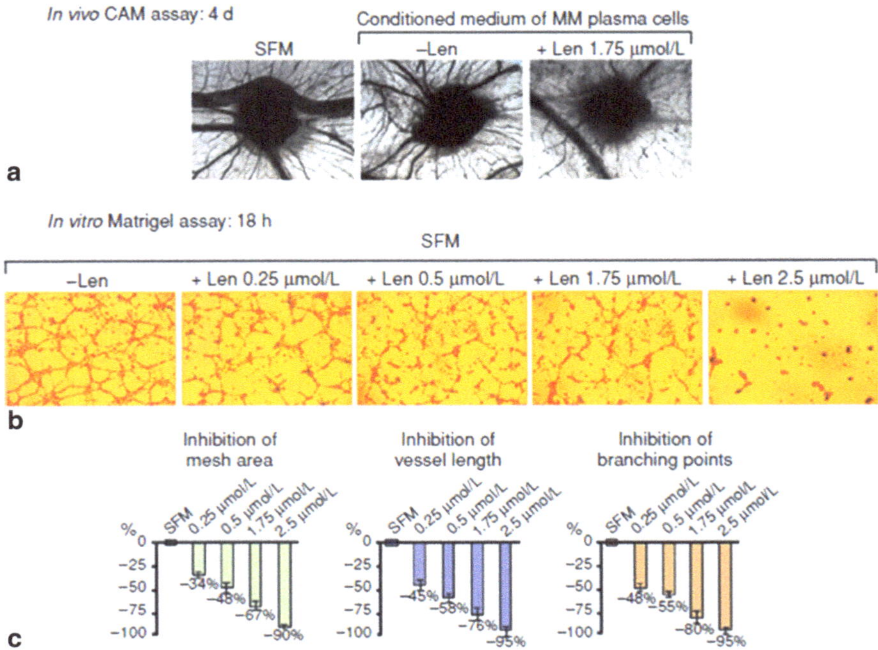

Fig. 5.1 Lenalidomide inhibits angiogenesis in CAM and Matrigel. **a** CAMs were incubated with gelatin sponges loaded with SFM (*left*) and with conditioned medium of multiple myeloma (MM) plasma cells either alone (*middle*) or supplemented with 1.75 mmol/L lenalidomide (*right*). Note the inhibition of MM angiogenesis by the drug. **b** lenalidomide inhibits multiple myeloma endothelial cells (MMEC) angiogenesis in the Matrigel in a dose-dependent manner. MMECs arranged to form a closely knit capillary-like plexus (*left*), whereas the tube formation was gradually blocked with increasing lenalidomide doses with a full inhibition at 2.5 mmol/L (*right*). A representative patient is shown. **c** skeletonization of the mesh was followed by measurements of its topological parameters: mesh area, vessel length, and branching points. Data are presented as mean ± SD of percent inhibition. Len, lenalidomide. (Reproduced from De Luisi et al. 2011)

et al. 2008). Moreover, thalidomide has shown antitumor activity in combination with rituximab in patients with relapsed or refractory MCL (Kaufmann et al. 2004).

In CLL, lenalidomide had considerable activity in both heavily pre-treated patients and patient with unfavorable prognostic factors (Ferrajolil et al. 2008; Chanan-Khan et al. 2006). In CLL, serious side effects of thalidomide and lenalidomide have been reported, including myelosuppression, tumor flare reaction and tumor lysis syndrome (Awan et al. 2010).

Thalidomide analogues appear to have significantly greater potency than thalidomide with a more favourable side effect profile, and these new agents have important implications for changing the current paradigm of anti-myeloma therapy. Continued study of thalidomide is warranted, especially with laboratory correlates to further elucidate its mechanisms of action and to determine the ideal dosing schedule, duration of therapy enrolment and overall maintenance therapy. Of special interest will be its role in combination with both standard agents and new drugs.

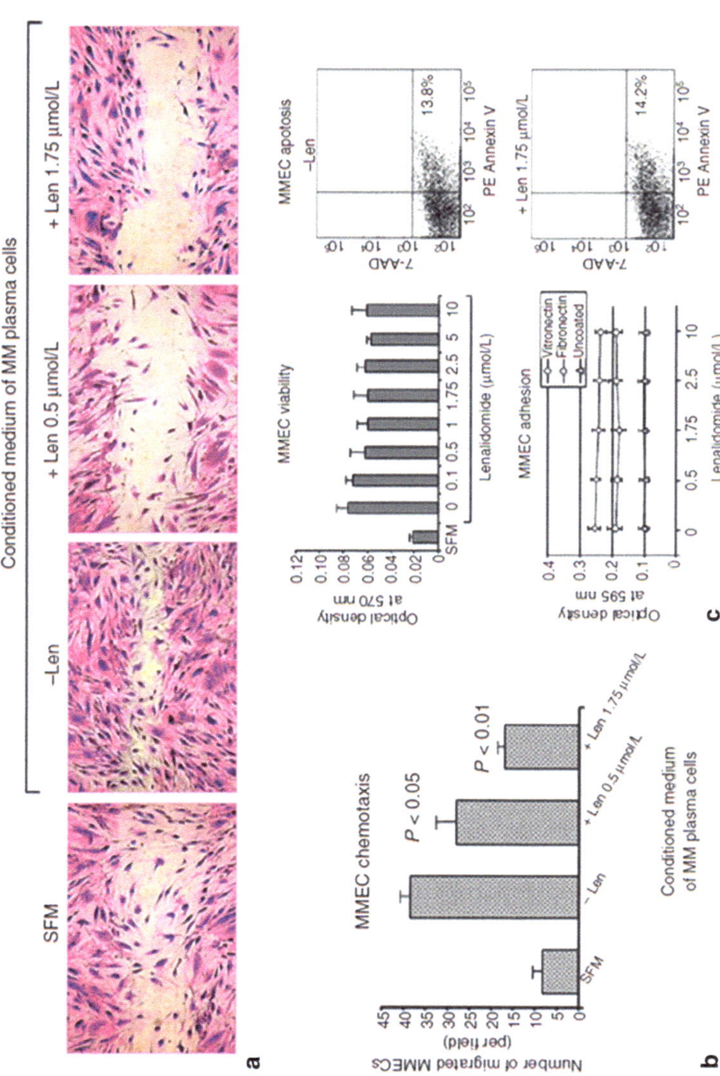

Fig. 5.2 Lenalidomide affects MMEC functions associated with angiogenesis. It inhibits MMEC migration in a dose-dependent manner, using the "wound" healing (**a**) and the Boyden microchamber (**b**) assays. **c** Lenalidomide increasing doses do not affect MMEC viability and adhesion, nor induce cell apoptosis. Len, lenalidomide. (Reproduced from De Luisi et al. 2011)

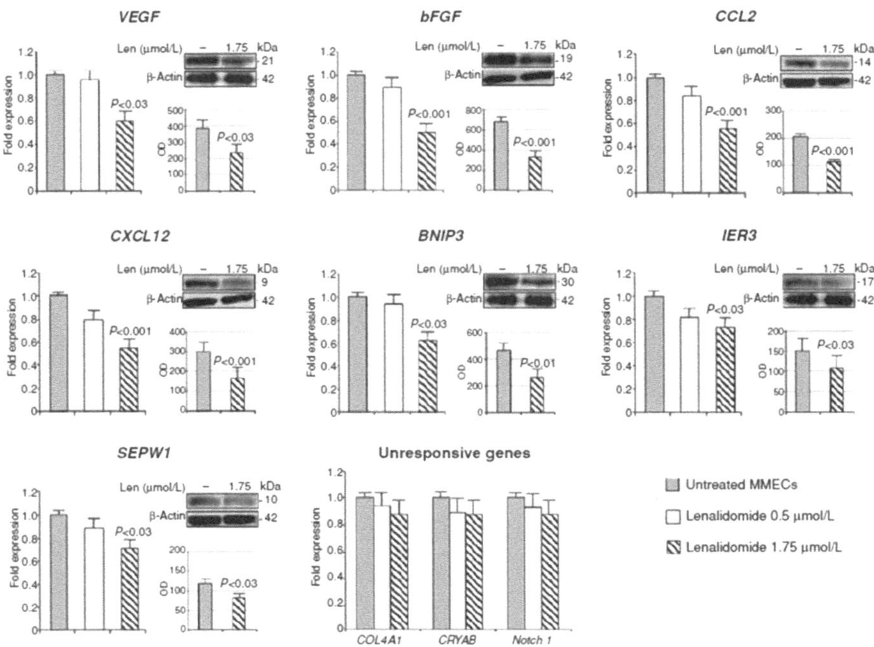

Fig. 5.3 Lenalidomide downregulates key genes and proteins of MM angiogenesis. Expression levels of genes (real-time RT-PCR) and proteins (Western blotting) in MMECs untreated and treated with lenalidomide. Len, lenalidomide. (Reproduced from De Luisi et al. 2011)

5.6 Combination Therapy

Thalidomide has also been combined with other active agents in the management of relapsed MM, while the absence of myelosuppression and other important adverse effects suggests that it could be combined with chemotherapy. Combination with other agents has produced higher overall response rates (defined as a greater than 50 % reduction in myeloma paraprotein levels) in relapsed and relapsed, refractory myeloma, with a mean of response rate of 36 % with thalidomide alone, compared to 52 % when combined with dexamethasone and 62 % when combined with dexamethasone and chemotherapy (Fanelli et al. 2003). Importantly, after myeloablative therapy for MM, it has been shown that progression-free survival is shorter for disease in partial remission as compared to complete remission (Fanelli et al. 2003). In an attempt to induce more durable remission, thalidomide plus dexamethasone has thus been employed in patients with partial remission after intensive therapy (Alexanian et al. 2002). The combination reduced tumor mass in 57 % of patients and this may have resulted in longer disease-free survival with preservation of tumor sensitivity to treatment with previously effective drugs also suggested in this study (Alexanian et al. 2002). When given within 15 months after intensive therapy, thalidomide plus dexamethasone, provided myeloma protein production has been reduced by 475 %, kept it at a constant level for at least 4 months, and markedly reduced the tumor mass in 57 % of patients, with stable, residual disease. The combination has also been evaluated for the primary management of active MM (Dimopuolos et al. 2001a) and untreated symptomatic forms (Weber et al. 2003) and appears especially active in the context of a completed Eastern Cooperative Oncology Group (ECOG) trial (Sparano et al. 2004). Palumbo et al. (2004) compared the clinical outcome of patients with relapsed/refractory MM with that of a control group treated with conventional chemotherapy and demonstrated that as first salvage regimen, thalidomide–dexamethasone was superior to conventional chemotherapy, while as second or third salvage regimen, it was equivalent to conventional chemotherapy. Thalidomide plus cyclophosphamide and dexamethasone is active in 77 % of patients (Garcia-Sanz et al. 2002). The addition of etoposide increases this figure to 86 % (Moehler et al. 2001). Anthracyclines are associated with a greater risk of thalidomide-induced DVT (Zangari et al. 2001). Bortezomib (previously denoted PS-341) is a novel proteasome inhibitor recently approved by the US FDA for the therapy of patients with progressive MM after previous treatment (Kane et al. 2003; Richardson et al. 2003). This drug has also been approved by the European Agency for the evaluation of Medicinal Products (EMEA) and most recently with the results of the APEX study has been shown to be effective in the treatment of first relapse. Importantly, bortezomib alone or in combination with dexamethasone and thalidomide has since demonstrated remarkable therapeutic activity in heavily pretreated patients with advanced MM including patient who have had prior thalidomide and bortezomib as single agents (Jagannath et al. 2004). This represents an important paradigm shift supporting the use of combinations of biological therapies for the treatment of MM patients.

Richardson et al. (2009, 2010) showed that lenalidomide, bortezomib, and dexametasone achieved 58 % response in relapsed-refractory MM and demonstrated that this combination therapy for newly diagnosed MM achieved 100 % responses, with 74 % at least very good partial and 52 % complete or near-complete response.

5.7 VEGF Neutralizing Antibodies

The VEGF signaling pathway can be inhibited at various levels, i.e. by blocking the VEGF activity with monoclonal antibodies (Jung et al. 2002), blocking the VEGFR with specific inhibitors (Kim et al. 2002) or interfering with the tyrosine kinases activated by the VEGF/VEGFR interaction (Wedge et al. 2002). Systemically administered antibodies to VEGF-A accumulate selectively in tumor vessels in much higher concentrations than in normal tissue (Ke-Lin et al. 1996). These antibodies and those that block VEGF-AR retard tumor growth and reduce tumor size in mice (Kim et al. 1993; Millauer et al. 1994). The latest antibodies selectively recognize the complex that VEGF-A forms with VEGFR-2 on vascular endothelium (Brekken et al. 2000). Preclinical studies with VEGF-neutralizing antibodies had shown that inhibition of VEGF inhibits MMP production and induction of apoptosis in cells expressing VEGFR (Presta et al. 1997), suppresses the production of TNF-α, and IL-1α in bone marrow stromal cells and inhibits leukemia colony formation in cultures of cells from patients with CML and refractory anaemia with blast excess (Bellamy et al. 2001a).

Aplidine is not an antibody, but a marine natural product used as an anti-cancer agent, that inhibited tumor growth and induced tumor cell apoptosis in human leukemia cells through inhibition of VEGF secretion and blockade of VEGF/VEGFR-1 autocrine loop (Broggini et al. 2003).

Antibodies to VEGF or expression of a dominant-negative VEGF receptor inhibit tumor growth *in vivo* without affecting tumor cell proliferation. In animals treated with anti-VEGF antibodies, the density of blood vessels in tumor sections is lower than in the tumors of control a mediated by blockade of VEGF's angiogenic activity. A neutralizing anti-VEGFR-2 antibody inhibiting 50 % of AML cell proliferation *in vitro* prolonged the survival of mice xenografted with human leukemia (Dias et al. 2000), supporting the notion that angiogenesis is important in the pathology of leukemia as well as solid tumors.

A humanized murine anti-VEGF monoclonal antibody, bevacizumab (Avastin), which recognizes all VEGF-A isoforms without cross-reacting with other growth factors, has displayed potent antitumor activity in experimental models (Chen et al. 2001).

Bevacizumab inhibit tumor growth, either alone or in combination with chemotherapy in untreated DLBCL (Ganjoo et al. 2006; Stopeck et al. 2009). A long disease-free survival in patients with aggressive NHL subtypes treated with bevacizumab as single agent has been reported (Stopeck et al. 2009). Anti-VEGF neutralizing antibodies and VEGFR inhibitors blocked the pro-survival effect of CD154 (CD40 ligand) on CLL cells and decreased the migration of CLL cells through the endothelium (Farahoani et al. 2005; Till et al. 2005).

The single agent therapy with bevacizumab did not show effective clinical activity in 13 patients with relapsed-refractory CLL in phase II trials (Shanafelt et al. 2010), and in adult patients with AML after chemotherapy (Karp et al. 2004; Zahiragic et al. 2007). Bevacizumab significantly caused decrease of VEGF expression without changes in VEGFR-2 expression and phosphorylation in bone marrow of AML patients (Zahiragic et al. 2007). In a phase II clinical trial bevacizumab was used in patients with either relapsed or refractory AML, accordingly to this schedule: cytarabine was administered on day 1–4, followed by mitoxantrone, and finally bevacizumab on day 8 (Karp et al. 2004). The results of this study demonstrated that serum VEGF levels elevated prior to bevacizumab infusion, decreased markedly 2 h after. Moreover, overall response was 48 % with complete response in 33 %. Median overall and disease-free survivals of complete response patients was 16.2 months and, respectively, 7 months. Bone marrow samples demonstrated marked MVD decrease after bevacizumab administration (Zahiragic et al. 2007). Mesters et al. (2007) assessed the activity of single agent bevacizumab in 9 patients with relapsed and refractory AML and demonstrated that VEGF expression in the bone marrow was decreased, while VEGFR-2 and blast count were not decreased.

Anti-VEGF neutralizing antibodies and VEGFRs inhibitors blocked the pro-survival effect of CD154 (CD 40 ligand) on CLL cells and decreased the migration of CLL cells through the endothelium (Till et al. 2005; Ellis et al. 2008).

5.8 Receptor Tyrosine Kinase Inhibitors

Receptor tyrosine kinases (RTKs) are transmembrane proteins containing an extracellular lectin binding domain and an intracellular catalytic domain. Many of the processes involved in tumor growth, progression and metastasis are mediated by signaling molecules acting downstream from activated RTKs. Tyrosine kinase inhibitors (TKIs) are small molecules able to pass the plasma membrane (Imai and Takaoka 2006). The tyrosine kinase VEGFRs are crucial mediators in angiogenesis and stimulation of VEGFRs and other RTKs causes massive activation of signaling pathways in endothelial cells. Tyrosine kinase inhibitors inhibit not only VEGFRs but also other receptors in the superfamily of RTKs, including the PDGFR. Inhibitors of VEGF signaling not only interfere with angiogenesis but also cause regression of some tumor vessels (Bergers et al. 2003), causing changes in all components of the vessel wall of tumor, consisting in loss of endothelial cell fenestrations, regression of tumor vessels, and appearance of basement membrane ghosts (Inai et al. 2004). In 2005, the FDA granted regular marketing approval for sorafenib, a small and oral inhibitor for the treatment of patients with advanced renal cell carcinoma (Kane et al. 2006a). Sunitinib is a small oral multikinase inhibitor of VEGFR, PDGFR, *c-kit* and Flt-3 kinase activity (Cabebe and Wakelee 2006). Tyrosine kinase inhibitors can be taken orally, if necessary in a salt form of the inhibitor. For example, sunitinib is taken as sunitinib malate, while sorafenib as tosylate sorafenib. Tyrosine kinase inhibitors can be subdivided in three categories. Type I kinase inhibitors recognize the active conformation of a kinase. An example is sunitinib, which demonstrates competitive inhibition to adenosine triphosphate (ATP) agonist VEGFR-2 and PDGFR-β. Type II kinase inhibitors recognize the inactive conformation of a kinase. An example is sorafenib, which blocks the phosphorylation of VEGFR, PDGFR, *Raf* and *kit* by using a hydrophobic packet to indirectly compete with ATP. A third class of kinase inhibitors is known as "covalent" inhibitors and have been developed to covalently bind to cysteines at specific sites of the kinases. An example is vandetanib, which in addition to targeting VEGFR, inhibits epidermal growth factor receptor (EGFR) (Morabito et al. 2009). An advance in this field includes the development of soluble decoy receptor incorporating both VEGFR-1 and VEGFR-2 domains (VEGF-Trap), binding VEGF with higher affinity than previously reported VEGF antagonists (Holash et al. 2002). The VEGF-Trap abolished mature, pre-existing vasculature in established xenografts resulting in almost completely avascular tumors subsequently followed by marked tumor regression and suppressed tumor growth (Holash et al. 2002).

5.8.1 Receptor Tyrosine Kinase Inhibitors in the Treatment of Multiple Myleoma

Lin et al. (2002) showed that vatalanib (PTK787/ZK222584), an orally administered broad-spectrum TKI of VEGFR-1, -2, -3, PDGFR-β, *c-kit*, inhibited proliferation and migration of MM cells. Pandiella et al. (2003) showed that imatinib mesylate (STI 571) blocked cell-cycle progression in MM and potentiated the effects of conventional antimyeloma agents *in vitro*. However, in a phase II clinical trial in patients with refractory/relapsed myeloma, no response was obtained (Dispemzieri et al. 2006). Zangari et al. (2004) and Kovacs et al. (2006) evalutated the activity of, respectively, SU5416 a small TKI of VEGFR-1, -2, -3 and of vandetanib (ZD6474) in patients with refractory MM and observed a decrease in VEGF serum levels in patients with stable disease, but not objective response. Podar et al. (2004, 2006) demonstrated that pazopanib (GW786034B) and GW654652, two broad-spectrum TKIs of VEGFR-1, -2, -3, PDGFR, *c-kit*, inhibited *in vivo* MM cell proliferation, migration and survival, VEGF-induced up-regulation of adhesion molecules on both endothelial and tumor cells, and exerted an anti-angiogenic activity *in vivo*. However, a recent phase II clinical trial in 21 MM patients treated with pazopanib, did not show any clinical response (Prince et al. 2009). Ramakrishnan et al. (2010) showed that sorafenib exerted a significant anti-myeloma activity and synergized with common anti-myeloma drugs. Coluccia et al. (2008) has shown constitutive activation of PDGFR-β/*Src,* two dasatinib targets, in plasma cells and endothelial cells isolated from patients with MM. Moreover, dasatinib significantly delayed MM tumor growth and angiogenesis *in vivo* (Fig. 5.4) showing a synergistic cytotoxicity with other anti-myeloma drugs, *i.e.* melphalan, prednisone, bortezomib, and thalidomide. In about 10–20% of MM patients, a translocation [t(4;14)] involving FGF receptor 3 (FGFR-3) is associated with poor prognosis (Avet-Loiseau et al. 1998; Keats et al. 2003; Qing et al. 2009). Small molecules with selective TKI activity (SU5402, SU10991, PD173074, PKC412) have been validated in preclinical models of MM (Grand et al. 2004; Paterson et al. 2004; Trudel et al. 2004).

As observed in patients with solid tumors, the most consistent side effect is hypertension, which is usually manageable with medical therapy. Other serious side effects include thromboembolic events, ischemic cerebrovascular accidents and congestive heart failure. Furthermore, bleeding complications and wound healing problems may be caused by a disturbance of the interaction of platelets with the vasculature.

5.8.2 Receptor Tyrosine kinase Inhibitors in the Treatment of Leukemia

Paesler et al. (2010) demonstrated that vatalanib and pazopanib decreased phosphorylation of the VEGFR of primary CLL cells with induction of apoptosis. Moreover, when combined with cytotoxic agents (fludarabine, vincristine and doxorubicin), vatalanib and pazopanib significantly decreased CLL cells survival rate compared to cytotoxic agents alone. In a model of xenografted mice, both vatalanib and pazopanib inhibited tumor growth. Vatalanib is well tolerated in AML patients with minimal side effects (Roboz et al. 2006). There was no significant response to vatalanib treatment in refractory/relapsed AML patients upon monotherapy, while a subgroup of patients with secondary AML responded to vatalanib in combination with chemotherapy. Barbarroja et al. (2009) utilized vatalanib in combination with idarubicin in 4 AML cell lines and 7 AML patient samples and demonstrated that vatalanib decreased VEGF levels and VEGFR phosphorylation in AML cells.

Cediranib is used in the treatment of glioblastomas, non small cell lung cancer (NSCLC), gastro intestinal stromal tumors (GIST) and renal cell carcinoma (Wedge et al. 2005). A correlation has been established between cediranib exposure

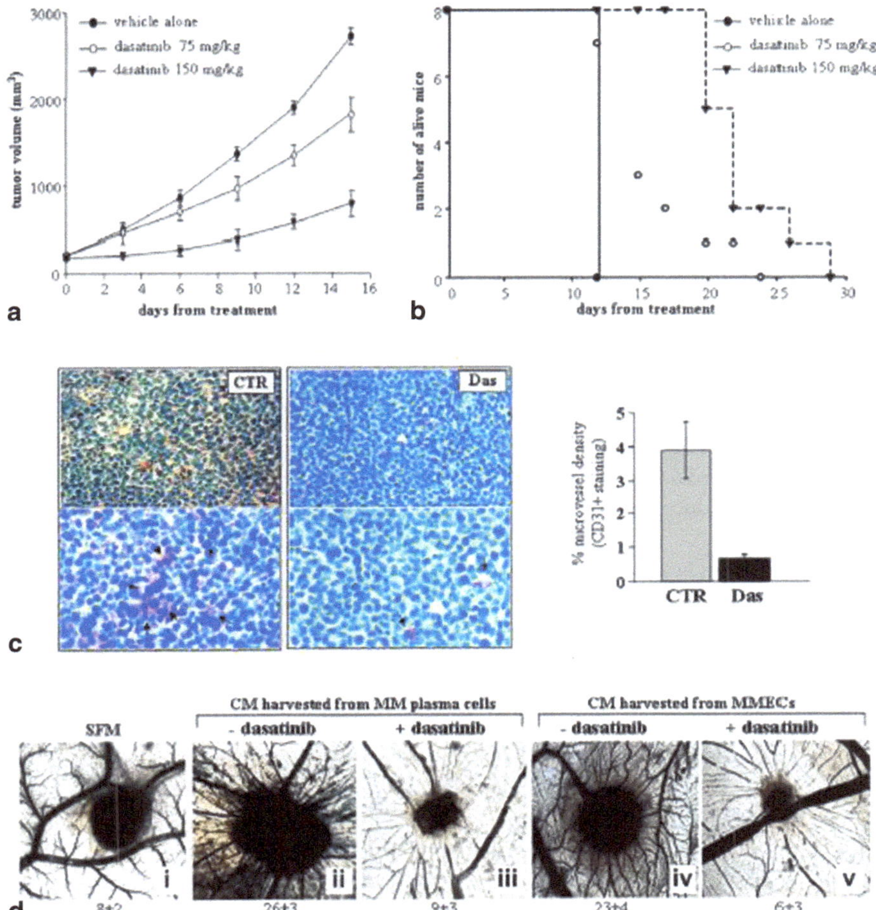

Fig. 5.4 Dasatinib inhibits multiple myeloma (MM) tumor growth and angiogenesis *in vivo*. (**a**) Animals were killed when their tumors reached 2 cm in diameter or when paralysis or major compromise in their life occurred. Tumor volume is indicated as mean. (**b**) Survival was evaluated from the first day of treatment until death. (**c**) Representative microscopic images of tumor sections costained with hematoxylin and eosin and anti-CD31. (**d**) Angiogenic responses induced by gelatin sponges soaked with MM patient plasma cells (*ii*) or endothelial cells (*iv*) conditioned-medium compared with the sponge treated with medium alone (*i*). After treatment of the CAM with 50 nM dasatinib (*iii* and *v*, respectively), significantly fewer vessels surrounded the sponges. Vessel counts are indicated. (Reproduced from Coluccia et al. 2008)

and plasma VEGF levels and reduction of soluble VEGFR-2 in a phase I clinical trial in AML, associated to a significant decrease in bone marrow MVD, while modest benefit was reported in clinical response (Fiedler et al. 2010).

Sunitinib inhibits tumor angiogenesis with some antiproliferative and apoptotic effects, as it has been demonstrated in murine xenograft models (Mendel et al. 2003). In AML tumor cells sunitinib decreased VEGF production and inhibited FLT3 phosphorylation (O'Farrell et al. 2003). In AML patients, sunitinib showed a slight effect (none of the patients showed a complete remission) in a phase I clinical trial used a single agent (Fiedler et al. 2005; O'Farrell et al. 2003a). Sunitinib has further inhibitory effects on AML FLT3 cell survival when combined with cytarabine or daunorubicin (Yee et al. 2004). The capability of blockade sunitinib to inhibit the growth and to induce apoptosis in AML cells was potentiated by blockade of MEK/ERK signaling (Nishioka et al. 2007).

Axitinib inhibits angiogenesis and induces tumor cell apoptosis (Hu-Lowe et al. 2008). Axitinib was used in a phase II clinical trial in the treatment of elderly AML patients with poor prognosis, without clinical responsiveness (Giles et al. 2006).

Midostaurin exerts antiproliferative activity against various tumor cell lines (Fabbro et al. 2000; Stone et al. 2005, 2012). A phase II clinical trial was undertaken in advanced AML patients with scarce benefits (Stone et al. 2005). New diagnosed AML patients were treated with midostaurin concurrently with induction therapy and the complete remission rate for the midostaurin 50 mg-twice-daily dose schedule was 80% (Stone et al. 2012). Midostaurin demonstrated a high percentage of partial responses in AML patients in a monotherapy phase II clinical trial (Fisher et al. 2010).

Vandetanib induces growth arrest and apoptosis in several myeloid keukemia cell lines (Nishioka et al. 2007; Jia et al. 2009). SU5614 inhibits VEGF-induced endothelial cell sprouting, growth arrest and apoptosis and by inhibition of cKIT in AML cells (Sun et al. 1998; Spiekermann et al. 2002). Combination of SU5614 with cytotoxic agents have shown better effects compared with monotherapy (Aleskog et al. 2005).

Lestaurtinib has been tested with positive results in phase II studies in patients with refractory or relapsed or poor-risk AML (Levis et al. 2002; Smith et al. 2004; Knapper et al. 2006).

Sorafenib in pre-clinical models has shown antiproliferative and anti-angiogenic activity as well as pro-apoptotic activity in tumor cell lines (Dal Lago et al. 2008). In 2005, the FDA granted regular marketing approval for sorafenib for the treatment of patients with advanced renal cell carcinoma (Kane et al. 2006). Sorafenib has been tested in AML patients (Ravandi et al. 2010; Metzelder et al. 2010). When administered in 8 AML patients FLT3 mutant, all patients showed rapid hematologic responses and 3 patients showed a complete molecular remission (Metzelder et al. 2010). However, a meta-analysis on AML patients in which oral sorafenib was administered versus placebo in combination with standard induction chemotherapy did not show improvements in event free survival and overall survival compared to the placebo group (Rollig et al. 2012).

Semaxinib produces a dose-dependent inhibition of tumor growth in a variety of xenograft models (Yee et al. 2002; Fong et al. 1999). *In vitro* studies on blasts from AML patients have shown a down-regulation of the expression of VEGF and signal transduction intermediated, such as STAT5 and Akt, after treatment with semaxinib (Loges et al. 2006). Semaxinib monotherapy led to a stable remission in a patient with a secondary AML relapse refractory towards conventional chemotherapy (Mesters et al. 2001), while a phase II clinical trial in patients with refractory AML demonstrated a partial response in 19% patients (Fiedler et al. 2003).

It is important to note that thalidomide, lenalidomide, anti-angiogenic agents and TKIs have all both antileukemic and immunomodulatory effects. This may be important id these drugus are used in allotransplanted patients (e.g., in the treatment of relapse after transplantation) (Reikvam et al. 2013).

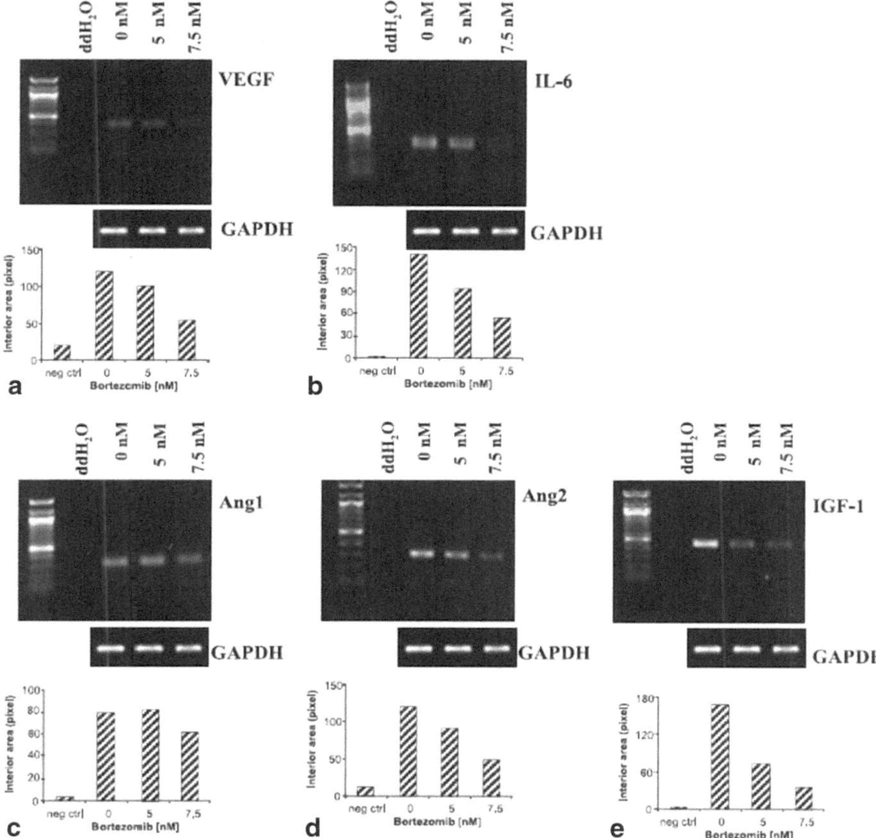

Fig. 5.5 Bortezomib down-regulates the expression of genes involved in angiogenesis. MMEC express VEGF, IL-6, IGF-1, Ang1,and Ang2. Bortezomib induced a down-regulation of all these genes in MMEC in a dose-dependent fashion. (Reproduced from Roccaro et al. 2006)

5.9 Bortezomib

Both at gene transcript and proteasome activity level, the ubiquitin proteasome cascade is up-regulated in MM, contributing to its enhanced activity. Bortezomib (Velcade, formerly PS-341) represents the first proteasome inhibitor to enter clinical trials. It induces endothelial cell apoptosis (Williams et al. 2003), inhibits VEGF, IL-6, Ang-1, Ang-2 and IGF-1 secretion in bone marrow stromal cells and endothelial cells derived from patients with myeloma (Roccaro et al. 2006; Hideshima et al. 2003a), HIF-1α activity (Shin et al. 2008), downregulates caveolin-1 tyrosine phosphorylation, which is required for VEGF-mediated MM cell migration, thereby inhibiting ERK-dependent cell proliferation (Hideshima et al. 2003b). Roccaro et al. (2006) have demonstrated that bortezomib inhibits the proliferation of MM endothelial cells in a dose- and time-dependent manner and down-regulates the expression of genes involved in angiogenesis (Fig. 5.5). Moreover, in functional

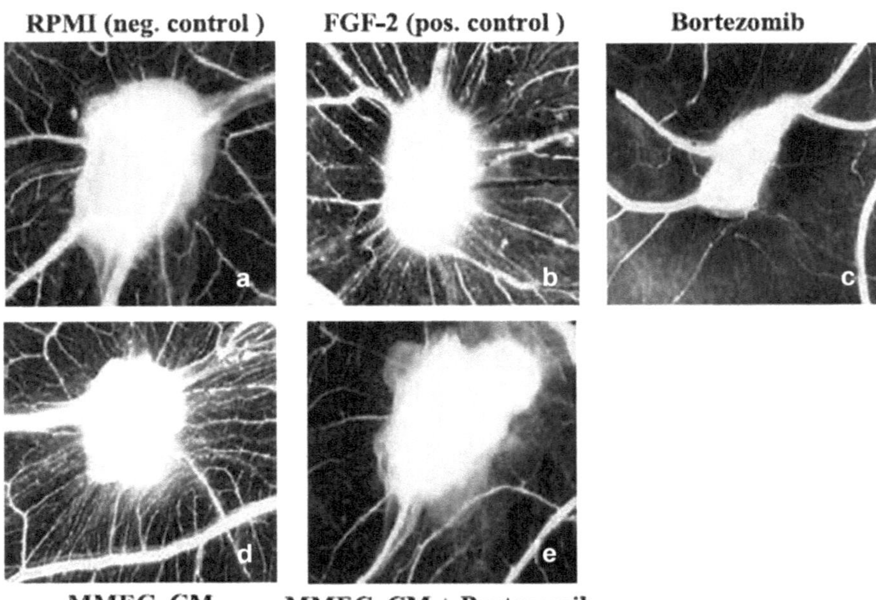

Fig. 5.6 Bortezomib inhibits angiogenesis *in vivo* in the chorioallantoic membrane (CAM) assay. The CAM was incubated with a gelatin sponge loaded with RPMI (**a**), 200 μg/mL FGF-2 (**b**), 20 nmol/L bortezomib (**c**), and with multiple myeloma endothelial cell (MMEC) conditioned media (CM) either alone (**d**) or with 20 nmol/L bortezomib (**e**). Bortezomib significantly inhibited basal angiogenesis induced by sponges loaded with vehicle alone (**c**). Moreover, CAM implanted with FGF-2 (**b**) or with MMEC conditioned media (**d**) increased macroscopic vessel counts. Bortezomib significantly inhibited MMEC CM–induced angiogenesis (**e**), evidenced both by macroscopic vessel counts and by the number of eggs out of total with > 50 % inhibition in the angiogenic response compared with vehicle. (Reproduced from Roccaro et al. 2006)

assay of angiogenesis, including chemotaxis, adhesion to fibronectin, capillary formation on Matrigel, and CAM assay, bortezomib demonstrated a dose dependent inhibition of angiogenesis (Fig. 5.6). Bortezomib has been previously approved for MM patients who failed at least one prior therapy (Kane et al. 2006b), and for initial treatment of patients with MM in a pivotal, multicenter, open-label trial, in which 682 previously untreated MM patients, who were ineligible for high-dose therapy plus stem-cell transplantation, were randomized to receive melphalan and prednisone combination alone (control group) or with bortezomib (San Miguel et al. 2008). The time to progression among patients receiving bortezomib plus melphalan–prednisone was 24.0 months, as compared with to 16.6 months among those receiving melphalan–prednisone alone. The overall survival and response rates were also better in the bortezomib group (San Miguel et al. 2008). The use of bortezomib in pre-transplant induction therapy revealed a higher response rate, compared to other induction regimens (Rajkumar and Sonneveld 2009).

Bortezomib in preclinical studies appears not only to have activity against MM cells but also to down-regulate protective interactions with bone marrow stromal cells

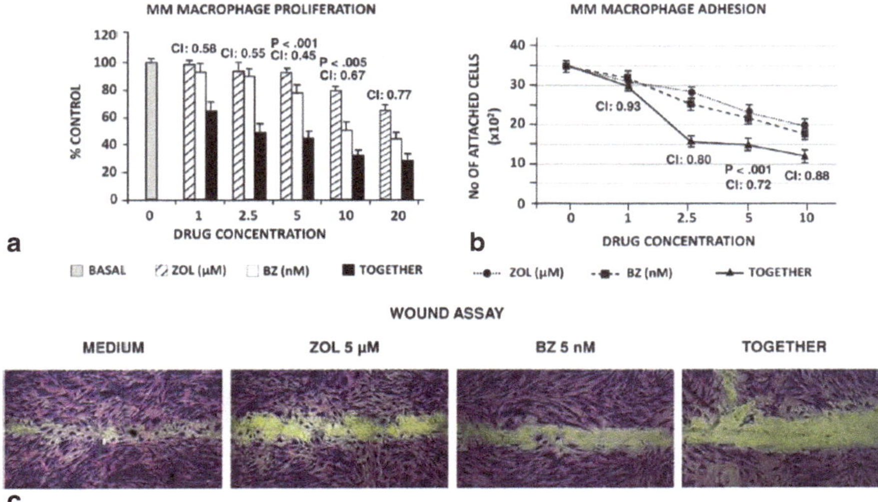

Fig. 5.7 Inhibitory effect of different doses of zoledronic acid (ZOL) and bortezomib (BZ), alone and together, on multiple myeloma (MM) bone marrow macrophages. Inhibition of (**a**) proliferation, (**b**) adhesion and (**c**) migration in the 'wound'. Synergistic inhibition is indicated by CI < 1. Significance of changes was assessed by the Wilcoxon signed-rank test. (Reproduced from Moschetta et al. 2010)

and to inhibit angiogenesis, and is now undergoing clinical evaluation in combination with thalidomide and CC-5013 for the treatment of MM patients in their earlier disease course (Anderson 2004).

We have demonstrated that bortezomib and zoledronic acid display distinct and synergistic activities on bone marrow macrophages in MM patients (Moschetta et al. 2010). They inhibited macrophage proliferation, adhesion, migration, expression of angiogenic cytokines (i.e. VEGF, FGF-2, HGF and PDGF), capillarogenesis on Matrigel, and VEGFR-2 and ERK1/2 phosphoactivation (Figs. 5.7, 5.8 and 5.9). Preclinical studies with new proteasome inhibitors are underway (Piva et al. 2008; Baumannn et al. 2009; Liu et al. 2012).

Clinical studies using bortezomib in relapsed or refractory B cell NHL, MCL, or marginal zone B cell lymphoma have shown promising results (Goy et al. 2005; Fisher et al. 2006; Belch et al. 2007; de Vos et al. 2009; O'Connor et al. 2010; Sehn et al. 2011).

5.10 Zoledronic Acid

The bisphosphonates pamidronate, zoledronic acid, and clodronate are employed for the treatment of patients with ostolytic lesions from MM. Bisphosphonates also may synergize with anticancer agents used in the treatment of myeloma, including dexamtehasone, thalidomide, and bortezomib.

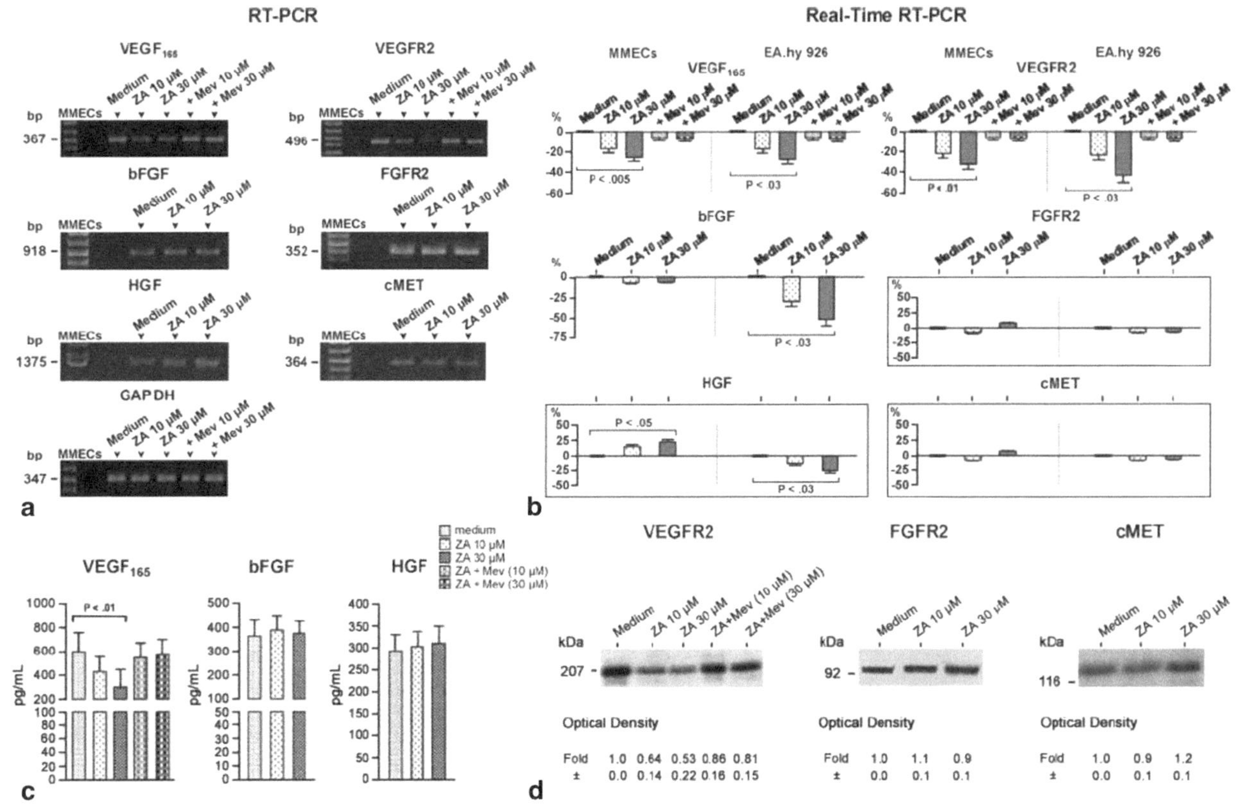

Fig. 5.8 Expression levels of VEGF, bFGF, HGF and PDGF genes in MM macrophages of a representative patient without (medium) and with exposure to ZOL and BZ, as evaluated by the (**a**) RT-PCR and (**b**) Real-Time RT-PCR. (**c**) ELISA analysis of the indicated cytokines secreted into the cell conditioned medium: synergistic inhibition is indicated by CI < 1. (Reproduced from Moschetta et al. 2010)

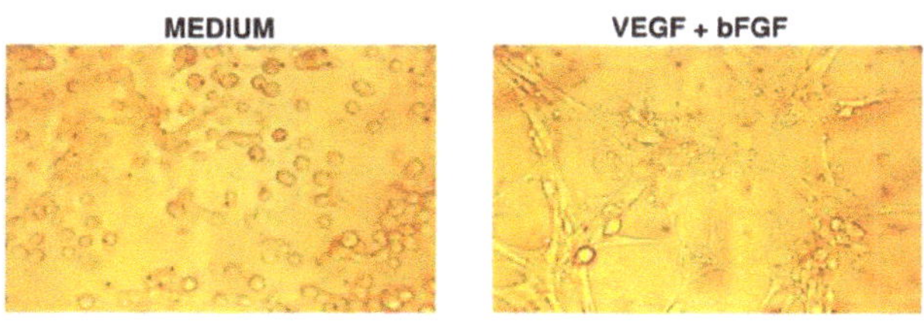

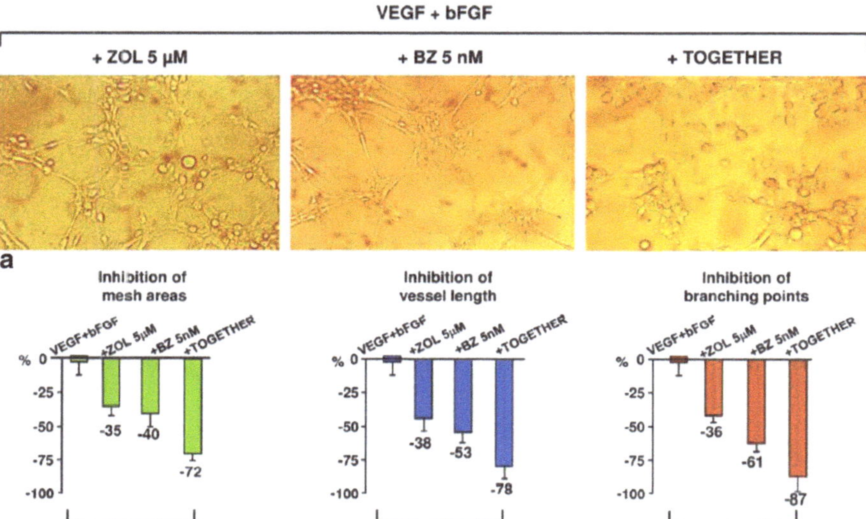

Fig. 5.9 Capillarogenesis on Matrigel. (**a**) Bone marrow vasculogenic macrophages of a representative patient were seeded on Matrigel and exposed to ZOL and BZ, alone and together. After a 24-h incubation, their 3-D organisation was examined planimetrically by computer image analysis. (**b**) Bars represent the mean ± 1 SD of percentage of inhibition of the indicated topological parameters. Synergistic inhibition is indicated by CI < 1. (Reproduced from Moschetta et al. 2010)

It has been established that zoledronic acid has a direct cytotoxic activity on tumor cells and suppresses angiogenesis (Clezardin 2002; Wood et al. 2002). We have demonstrated that therapeutic doses of zoledronic acid markedly inhibit *in vitro* proliferation, chemotaxis and capillarogenesis of MM endothelial cells and *in vivo* angiogenesis in the CAM assay (Fig. 5.10) (Scavelli et al. 2007). These effects are partly sustained by gene and protein inhibition of VEGF and VEGFR-2 in an autocrine loop. Mevastatin, a specific inhibitor of the mevalonate pathway, which prevents prenylation of several proteins leading to cellular apoptosis, anti-angiogenesis and activation of gamma/delta T-cells, reverts the zoledronic acid anti-angiogenic effect, indicating that the drug halts this pathway (Fig. 5.11).

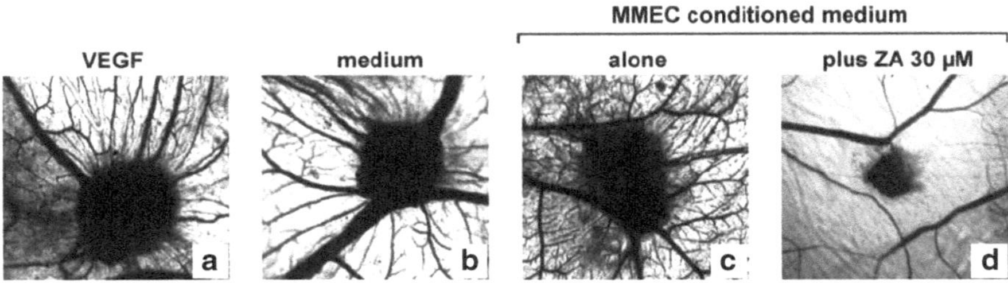

Fig. 5.10 Chorioallantoic membranes treated with sponges loaded with VEGF165 or multiple myeloma endothelial cells (MMEC) conditioned media were surrounded by allantoic vessels developing radially toward the implant in a spoked-wheel pattern (**a** and **c**). No vascular response was detectable around the sponges loaded with vehicle (medium) alone (**b**). Zoledronic acid added to the conditioned media significantly inhibits the angiogenic response (**d**). (Reproduced from Scavelli et al. 2007)

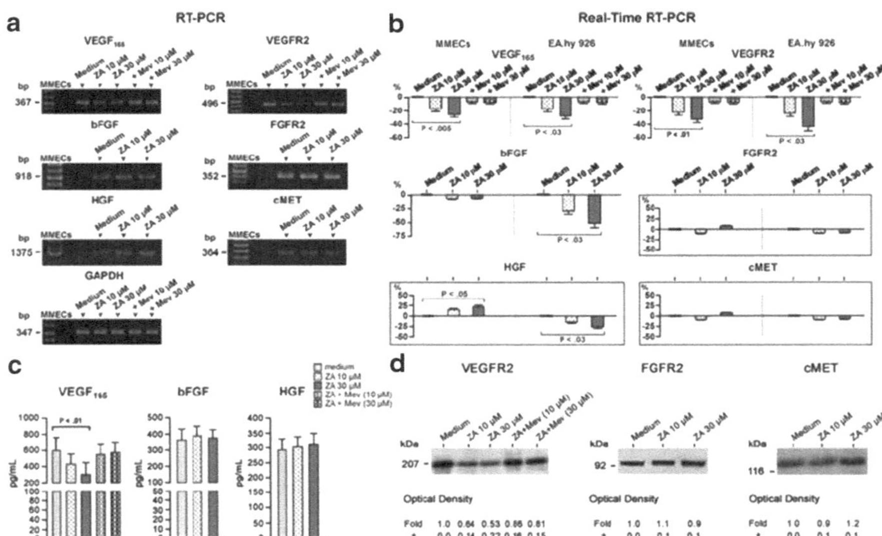

Fig. 5.11 **a** expression levels of VEGF165, VEGFR2, bFGF, FGFR2, HGF, and c-MET genes in MMEC after exposure to zoledronic acid alone or with mevastatin (Mev) at equimolar doses, as evaluated by RT-PCR. **b** real-time RT-PCR analyses on the same genes in MMECs and EA.hy926. Columns, mean percentage of inhibition or stimulation compared with the baseline value; **c** effect of zoledronic acid alone or added with mevastatin on VEGF, bFGF, and HGF concentrations in MMEC conditioned media as measured by ELISA. **d** protein levels of VEGFR2, FGFR2, and c-MET in MMEC lysates by Western blot analysis. (Reproduced from Scavelli et al. 2007)

Overall, these data suggest that the zoledronic acid antitumoral activity in MM is also sustained by anti-angiogenesis, which would partly account for its therapeutic efficacy in MM (Henk et al. 2012; Morgan et al. 2012).

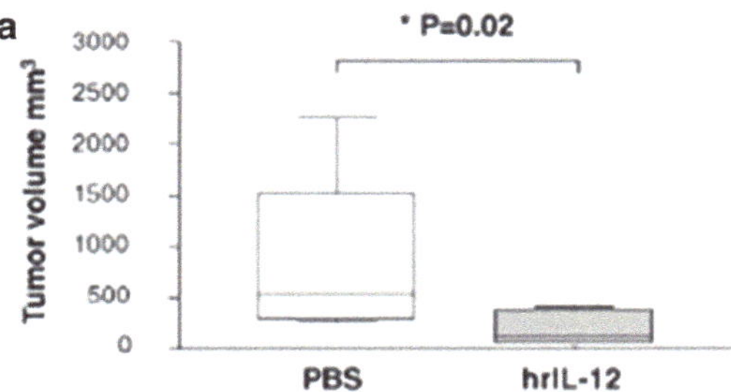

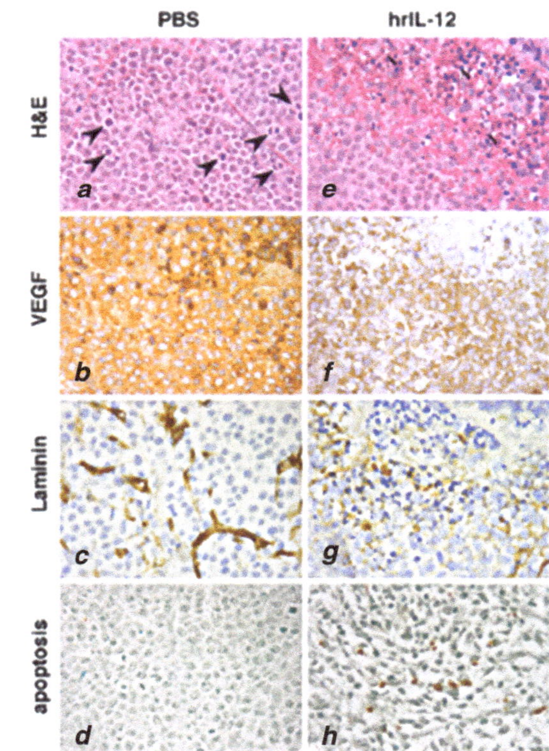

Fig. 5.12 Inhibition of U937 acute myeloid leukemia cell tumorigenicity in SCID–NOD mice by hrIL-12. **a** Volume of tumors grown i.p. in PBS or IL-12 treated mice 15 days after U937 cell injection. The differences in size between tumors removed from PBS and IL- 12 treated mice were evaluated by Mann–Whitney U test. Boxes indicate values between the 25th and 75th percentiles, whisker lines represent highest and lowest values for each group. Horizontal lines represent median values. **b** I.p. injection of human U937 cells in SCID–NOD mice gave rise to a small tumor mass (*a*) formed by highly proliferating (mitotic figures indicated by arrowheads) leukemic cells which expressed VEGF (*b*) and were supplied by a moderately developed vasculature, as revealed by laminin immunostaining (*c*). Apoptotic figures were almost absent (*d*). Treatment with hrIL-12 promoted ischemic–hemorragic necrosis (*e*) in association with a distinct suppression of VEGF expression (*f*) and a severe vascular deficiency (*g*). Apoptotic events were sometimes observed (*h*). (Reproduced from Ferretti et al. 2010)

5.11 Interleukins

We have investigated the ability of IL-12 and IL-27 in the regulation of human leukemias (Fig. 5.12) (Ferretti et al. 2010; Airoldi et al. 2006), lymphoma (Fig. 5.13) (Airoldi et al. 2008), and MM (Cocco et al. 2010a; Airoldi et al. 2004) cell growth. Supernatants from primary MM cells exposed to IL-12 showed a clear reduction of the angiogenic potential in comparison with untreated cells in the CAM assay. A PCR array performed in the same cells revealed that IL-12 treatment signifi-

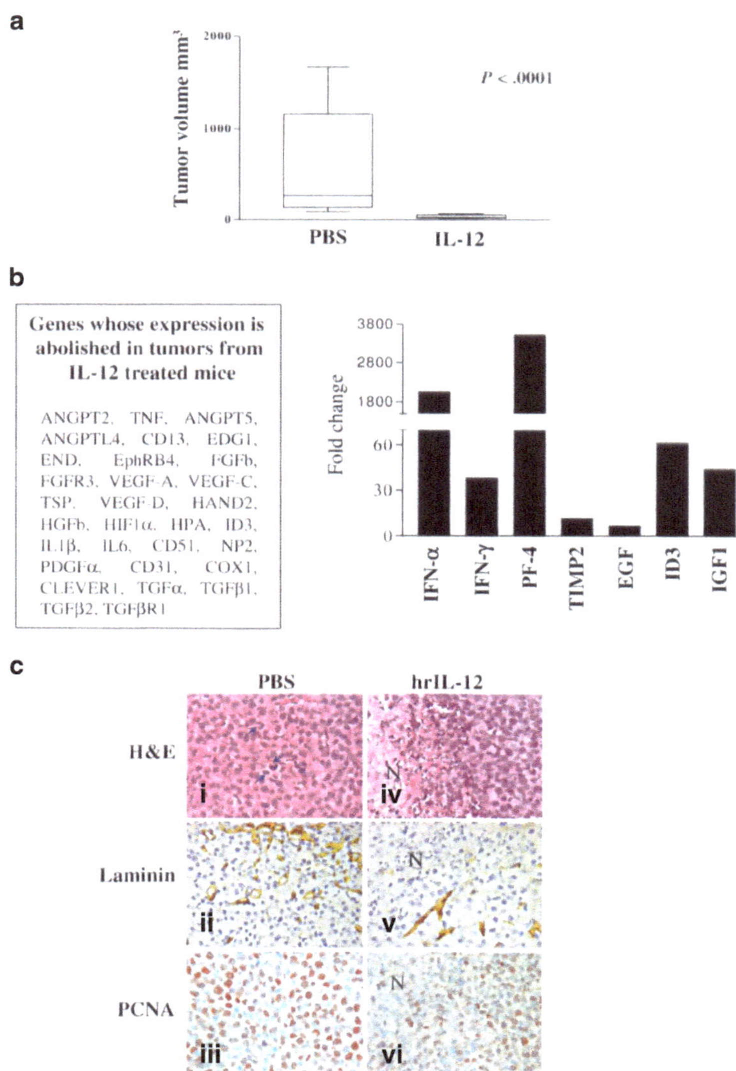

Fig. 5.13 IL-12 antitumor activity *in vivo* in multiple myeloma. **a** Volume of tumors grown intraperitoneally in PBS- and IL-12–treated animals 23 days after NCI-H929 human multiple myeloma cell inoculation. The differences in size between tumors removed from PBS and IL-12–treated mice were evaluated by Mann–Whitney U test. Boxes represent values between the 25th and75th percentiles; whisker lines, highest and lowest values for each group; horizontal lines, median values. **b** Human angiogenesis PCR array on tumors explanted from IL-12– versus PBS-treated animals 23 days after NCI-H929 cell inoculation. Left panel enlists the gene whose expression has been abolished in tumors from IL-12–versus PBS-treated mice. Histogram in right panel shows fold expression changes of genes up-regulated in tumors from IL-12– versus PBS-treated mice. **c** Histologic and immunohistochemical features of tumors developed in PBS-treated (*i–iii*) and hrIL-12–treated (*iv–vi*) SCID/NOD mice 23 days after NCI-H929 tumor cell injection. NCI-H929

cantly down-regulated mRNA of the proangiogenic CCL-11, whereas it up-regulated mRNA of the angiostatic chemokines CXCL-9 and CXCL-10, as well as the angiogenesis inhibitor IFN-γ (Cocco et al. 2010). In addition, *in vivo* studies showed that mice injected with the NCI-H929 or U266 MM cell lines and treated with IL-12 developed tumors significantly smaller than those treated with PBS. Morphological and immunohistochemical studies demonstrated that tumors from IL-12-treated animals displayed a wide focus of ischemic-coagulative necrosis, and reduced MVD, as assessed by laminin staining in comparison with control tumors. Molecular analysis highlighted that a wide panel of angiogenesis activators, including Ang-2, FGF-2, VEGF, IL-1ß, and IL-6, expressed during tumor neovascularization, was down-regulated by IL-12. Conversely, inhibitors of angiogenesis, including CXCL-4, IFN-α, IFN-γ, and TIMP-2, were found to be up-regulated in tumors grown in IL-12 versus PBS-treated animals (Cocco et al. 2010).

Human lymphoid malignancies may be targeted by the direct antitumor activity of each cytokine that functions by modulating the cellular angiogenic program. IL-12 exerts anti-tumor activity in the s.c. animal model injected with AML cells from a patient at diagnosis (Fig. 5.14) (Ferretti et al. 2012) and IL-12 treatment reduce the ability of human pediatric AML cells to induce new vessel formation in the CAM assay (Airoldi et al. 2006). These findings were supported by the *in vivo* experiments performed in severe combined immunodeficiency (SCID)-NOD mice injected s.c. with human U937 cells and subsequently, treated or not with human rIL-12. IL-12-treated animals displayed reduced tumor cell growth by increased apoptosis, decreased tumor cell proliferation, and inhibition of angiogenesis, as assessed by laminin immunostaining of blood vessels (Fig. 5.12) (Ferretti et al. 2010). These data have been confirmed in SCID-NOD mice injected s.c. with human acute lymphoblastic leukemia cells treated or not with human rIL-23 (Cocco et al. 2010b).

IL-27 directly inhibits MM cell growth both *in vitro* and *in vivo* primarily through the inhibition of angiogenesis (Cocco et al. 2010a; Giuliani et al. 2011). Moreover, IL-27 inhibits osteoclasts differentiation and activity, and induces osteoblast proliferation and damp *in vivo* tumorigenicity of human MM cell line through inhibition of angiogenesis. These results open new perspective for MM therapy because IL-27 may block MM progression and metastatic bone resorption. IL-27 inhibits tumor growth and angiogenesis *in vivo* in the CAM assay against pediatric B acute lymphoblastic leukemia cells (Canale et al. 2011).

---

tumors are mostly formed by undifferentiated, proliferating (mitotic features indicated by *arrows*) blast cells that are large and pleomorphic and sometimes binucleated or endowed with very prominent nucleoli (*i*). These tumors are supplied by a distinct network of mature microvessels, as assessed by laminin staining (*ii*), and show frequent PCNA expression (*iii*). In IL-12–treated mice, these morphologic features are frequently altered by the appearance of ischemic and hemorrhagic foci of necrosis (N; iv) associated with defective microvascularization (*v*) and decreased tumor cell proliferation (*vi*). (Reproduced from Airoldi et al. 2008)

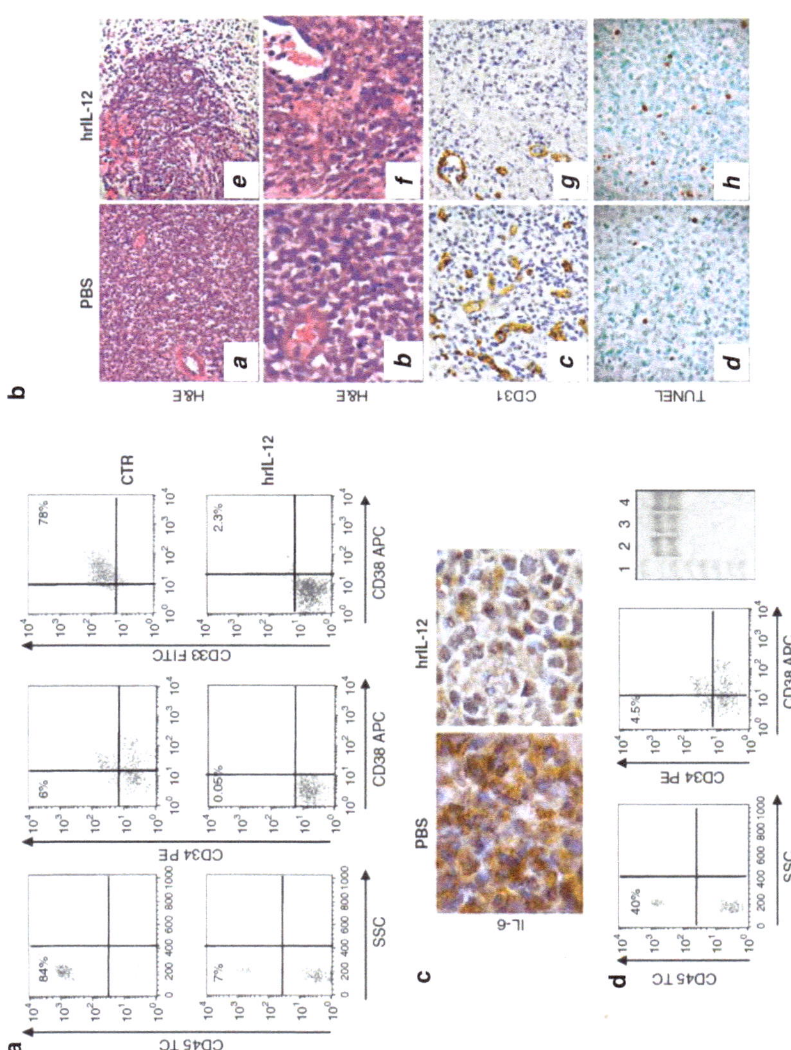

Fig. 5.14 IL-12 anti-tumor activity in the s.c. animal model. **a** Human CD45 positive (*left panels*), CD34 positive CD38 negative (*middle panels*), CD33 positive CD38 positive (*right panels*) acute myeloid leukemia cells found in tumors from control mice (*upper panels*) and hrIL-12-treated mice (*lower panels*)

5.12 Chemotherapeutics

Conventional cytotoxic chemoterapeutic drugs have been used at low (nanomolar) and noncytotoxic concentrations as anti-angiogenic agents (Vacca et al. 1999b). Anti-endothelial effects have been demonstrated *in vitro* for cyclophosphamide, 5-fluorouracil, and mitomycin C, and short-term anti-angiogenic effects have been demonstrated *in vivo* for vincristine, vinblastine, dexorubicin, mitaxantrone, etoposide, paclitaxel, 6-methylmercaptopurine, tegafur, 9-amino-20(s)-camptothecin, topotecan, camptosan, and combrestatine. The presence of dividing endothelial cells in newly forming tumor blood vessels should render them—in contrast to their mature, quiescent counterparts in normal adult tissues—sensitive to the cytotoxic affects of such drugs in a manner similar to that of dividing bone marrow cells (Kerbel et al. 2000). In this context, it has been demonstrated in murine models of leukemia that low-dose cyclophosphamide, used as an anti-angiogenic drug, may avoid drug resistance and eradicate established disease more effectively than the conventional maximum tolerable dose (Kerbel et al. 2000).

5.13 Histone Deacetylase Inhibitors and Vascular Disrupting Agents

Acetylation and deacetylation of histone proteins are important mechanisms for the regulation of gene expression. The interest in histone deacetylase (HDAC) as antineoplastic drugs originated with the observation that these agents could reverse

---

injected s.c. with acute myeloid leukemia cells from a patient at diagnosis. One representative experiment out of the four performed is shown. **b** Morphological and immunohistochemical features of tumors from animals injected s.c. with primary acute myeloid leukemia cells and treated with PBS (*a–d*) or hrIL-12. These tumors were analyzed for microvessel formation (CD31 staining: *c*, *g*), apoptosis (terminal deoxynucleotidyl transferase dUTP nick end labeling: *d*, *h*) and **c** IL-6 expression (tumors from control mice display diffuse expression of IL-6 whereas such expression is scanty in masses obtained from IL-12 treated mice). Tumors from controls are endowed with a finely branched microvasculature and rare apoptotic cells. Treatment with hrIL-12 heavily impaired the viability of most acute myeloid leukemia blasts and causes necrotic-hemorrhagic breakdown of the tumor masses which are crossed by few blood vessels. Moreover, these masses show more frequent apoptotic events than tumors from controls. **d** Human CD45 positive (*left panel*) and CD34 positive CD38 negative cells (*middle panel*) found in spleens of secondary recipients injected with acute myeloid lueukemia cells harvested from primary control recipients. Molecular analysis show that purified CD45 positive cells from the spleen of secondary recipients belong to the original acute myeloid leukemia clone. One representative experiment out of the four performed is shown. Right panel, gel molecular analysis of polymorphic short tandem repeat loci (Human FGA) in acute myeloid leukemia blasts from a patient at diagnosis (*lane 2*), in CD45 positive cells isolated from tumor mass of one representative primary recipient injected s.c. (*lane 3*) and in CD45 positive cells from the spleen of one representative secondary recipient injected s.c. with CD45 positive cells isolated from tumors grown in a primary control recipient (*lane 4*). Lane 1 shows the molecular weight markers. (Reproduced from Ferretti et al. 2012)

the malignant phenotype of transformed cells (Rasheed et al. 2008) HDAC represent an emerging class of therapeutic agents effective in haematological malignancies (Richon and O'Brien 2002) that induce tumor cells cytostasis, differentiation and apoptosis, in part due to an angiostatic effect, through an inhibition of VEGFRs expression in endothelial cells (Medina et al. 1997; Deaoanne et al. 2002).

Vorinostat, panobinostat, and MGCD0103 have been evaluated in NHL (Duvic et al. 2007; Crump et al. 2008). Successful therapy with vorinostat was associated with a reduced MVD and an increase of the anti-angiogenic molecule thrombospondin-1 following treatment (Deaoanne et al. 2002). Panobinostat was evaluated in CTCL and microarray analysis of skin biopsies showed a consistent down-regulation of pro-angiogenic gene guanylate cyclase 1A3 (GUCY1A3) and Ang-1 (Ellis et al. 2008). Two HDAC inhibitors, sodium butyrate and suberoylanilide hydroxamic acid reduced VEGF production and induced growth suppression and apoptosis in human mantle cell lymphoma cell lines (Heider et al. 2006).

Vascular disrupting agents (VDA) agents may be used in the treatment of leukemia (Madlambayan et al. 2010), and a marked survival benefit and therapeutic efficacy of two VDA, namely combrestatin-A1-diphosphate and the microtubule-targeting agent CA4P has been demonstrated in xenograft models of leukemia (Benezra et al. 2012; Petit et al. 2008).

Chapter 6
Concluding Remarks

The initial presumption that hematological malignancies are not dependent on neovascularization probably resulted from the lack of evidence that vascularity plays any role in development and growth of these tumors. After all, in most cases, the tumor cells are suspended in an hematopoietic environment that is replete with erythrocytes. However, we must keep in mind that these cells are often the progeny of malignant precursors, and that these precursors may be both adherent and localized to areas of the hematopoietic microenvironment that require vascularization for expansion. Thus, the fact that circulating leukemic cells have ready access to erythrocytes does not dismiss the possibility that their precursors, the cells that initiate the disease, do not. Similarly, MM eventuates in circulation of plasma cells that divide uncontrollably.

However, the first cells involved in the pathology of this disease may be cells adherent to stromal areas that require microvascularization for successful colonization. Maintenance of disease pathology may require the continued survival of tissue adherent cells, either to provide progenitors or growth factors needed to perpetuate the disease.

In lymphoma, the requirement for vascularization of affected lymph nodes is obvious. Thus, successful development of many hematological malignancies may well require localized neovascularization to provide nutrients and oxygen to the dysregulated blast cells that divide early in the course of the disease and continue to maintain the disease. Localization of the first cells involved in development of hematological malignancies may also account for the failure consistently to observe a generalized decreased in marrow vascularity associated with effective anti-angiogenic therapy for these diseases.

Thus, while decreased MVD following treatment with an angiogenesis inhibitor demonstrates that the agent is effective, the absence of such a decrease does not indicate that ineffective treatment. In addition to the requirement of localized neovascularization, if the tumor cells are eliminated rapidly, the thickness of the cuff of tumor cells encircling each capillary would be decreased and MVD increased, despite the fact that the tumor is regressing.

Thus, many factors may be involved in both the role of angiogenesis in hematological malignancies and the effects of therapy on the disease as well as the

D. Ribatti, *Angiogenesis and Anti-Angiogenesis in Hematological Malignancies*,
DOI 10.1007/978-94-017-8035-3_6, © Springer Science+Business Media Dordrecht 2014

hematopoietic vasculature. It is too early to predict whether anti-angiogenesis will be of wide clinical benefit in hematological malignancies. Strategies that target both the stromal and tumor compartments, such as combining traditional cytotoxic chemotherapy with anti-angiogenic agents, may markedly improve the therapeutic response.

The majority of clinical trials that target tumor angiogenesis use drugs that target the VEGF signaling pathway, either by blocking VEGF or interfering with VEGFR-2 signaling. Tumor vessels are immature and are characterized by an increased permeability. A normalization of tumor vasculature results as a consequence of anti-VEGF treatment leading to increased perfusion of the tumor and subsequent increase of oxygenation, as it has been demonstrated by morphological studies of tumor vessels and by intravital imaging studies. Nevertheless, anti-VEGF treatment promotes a pro-invasive phenotype and may even increase tumor metastasis (Ribatti 2011).

Anti-angiogenesis therapy faces some hurdles: intrinsic or acquired resistance, increased invasiveness and lack of biomarkers and reliable surrogate markers for therapeutic efficacy. Cancer cells are able to circumvent anti-angiogenic therapy and develop resistance to targeted monotherapy. Intrinsic resistance results from an absence of VEGF of VEGFR in certain tumors, as it has been demonstrated in pre-clinical studies (Karashima et al. 2007). Another mechanism of tumor resistance may be the consequence of co-option of existing blood vessels by tumor cells (Leenders et al. 2004). The development of hypoxia-resistant tumor subpopulations which can outgrowth the sensitive tumor cells or the selection of more mature, stable vessels that are intrinsically less responsive to anti-angiogenic treatment (Glade Bender et al. 2004), could also cause resistance. Moreover, tumors under hypoxic conditions, utilize compensatory mechanisms involving alternative pro-angiogenic pathways to elicit continued angiogenesis (Casanovas et al. 2005). Given the number of receptors, signaling pathways and molecules involved in angiogenesis, there is a potential redundancy among various pro-angiogenic factors and cross-talk within them and anti-angiogenic signaling pathways (Relf et al. 1997).

Multiple angiogenic molecules may be synthesized by tumor cells and tumors at different stages of development may depend on different factors for their blood supply. When a predominant pathway is suppressed, other pathways could become more active. Therefore, the anti-angiogenic therapy should disrupt the angiogenic process at multiple steps, using different compounds with different mechanism of action.

Another mechanism of resistance to the anti-angiogenic therapy may be the recruitment of bone marrow-derived EPCs, which can stimulate new-formation of tumors blood vessels (Lyden et al. 2001). It is important to note that anti-VEGF-VEGFRs therapies may cause a number of side effects, including diarrhea, leucopenia, and hypertension.

A clinical challenge is the finding of biological markers that help to identify subset of patients more likely to respond to a given anti-angiogenic therapy, to detect early clinical benefit or emerging resistances and to decide whether to change therapy in second-line treatment. Usefulness of markers of angiogenesis as prognostic markers in hematological malignancies seems reasonable although confirmation in larger patient cohort is required. Potential surrogate markers include serum, plasma and urine levels of VEGF and FGF-2, quantification of circulating EPCs.

In conclusion, despite to the exciting results obtained in preclinical studies, these have been not always confirmed in the clinical setting, in which different mechanisms of escape are active. It is unrealistic to expect a single agent to be efficacious in the treatment of leukemias. Anti-angiogenic agents will have to be used in combination with conventional agents or with other antiangiogenic drugs. In this context, angiogenesis inhibition prolongs progression-free survival, but has only small effects on overall survival, the remissions are partial or patients develop resistance to the treatment.

References

Aguayo AR, Estey E, Kantarajian H et al (1999) Cellular vascular endothelial growth factor is a predictor of outcome in patients with acute myeloid leukemia. Blood 94:3717–3721

Aguayo AR, Kantarajian H, Manshouri T et al (2000) Angiogenesis in acute and chronic leukemia and myelocysplastic syndromes. Blood 96:2240–2245

Aguayo A, Kantarjian HM, Estey EH et al (2002) Plasma vascular endothelial growth factor levels have prognostic significance in patients with acute myeloid leukemia but not in patients with myelodysplastic syndromes. Cancer 95:1923–1930

Aird WC (2009) Molecular heterogeneity of tumor endothelium. Cell Tissue Res 335: 271–281

Airoldi I, Di Carlo E, Banelli B et al (2004) The IL-12Rbeta2 gene functions as a tumor suppressor in human B cell malignancies. J Clin Invest 113:1651–1659

Airoldi I, Cocco C, Di Carlo E et al (2006) Methylation of the IL-12Rbeta2 gene as novel tumor escape mechanism for pediatric B-acute lymphoblastic leukemia cells. Cancer Res 66:3978–3980

Airoldi I, Cocco C, Giuliani N et al (2008) Constitutive expression of IL-12R beta 2 on human multiple myeloma cells delineates a novel therapeutic target. Blood 112:750–759

Aleskog A, Hoglung M, Pettersson J et al (2005) In vitro activity of the flt-3 inhibitor su5614 and standard cytotoxic agents in tumour cells from patients with wild type and mutated flt3 acute myeloid leukaemia. Leuk Res 29:1079–1081

Alexanian R, Weber D, Giralt S et al (2002) Consolidation therapy of multiple myeloma with thalidomide-dexamethasone after intensive chemotherapy. Ann Oncol 13:1116–1119

Anderson KC (2004) Clinical update: novel targets in multiple myeloma. Semin Oncol 31:27–32

Anderson AO, Anderson ND (1975) Studies on the structure and permeability of the microvasculature in normal lymph nodes. Am J Pathol 80:387–418

Andersen NF, Standal T, Nielsen JL et al (2005) Syndecan-1 and angiogenic cytokines in multiple myeloma: correlation with bone marrow angiogenesis and survival. Br J Haematol 128:210–217

Aref S, Mabed M, Zalata K et al (2004) The interplay between c-Myc oncogene expression and circulating vascular endothelial growth factor (sVEGF), its antagonist receptor, solube Flt-1 in diffuse large B cell lymphoma (DLBCL): relationship to patient outcome. Leuk Lymphoma 5: 499–506

Arias V, Soares FA (2000) Vascular density (tumor angiogenesis) in non Hodgkin's lymphomas and florid follicular hyperplasia: a morphometric study. Leuk Lymphoma 40:157–166

Asosingh K, De Raeve H, Menu E et al (2004) Angiogenic switch during 5T2MM murine myeloma tumorigenesis: role of CD45 heterogeneity. Blood 103:3131–3137

Ausprunk DH, Folkman J (1977) Migration and proliferation of endothelial cells in preformed and newly formed blood vessels during tumor angiogenesis. Microvasc Res 14:53–65

Avet-Loiseau H, Li JY, Facon T et al (1998) High incidence of translocations t(11;14)(q13;q32) and t(4;14)(p16;q32) in patients with plasma cell malignancies. Cancer Res 58:5640–5645

Awan FT, Johnson AJ, Lapalombella R et al (2010) Thalidomide and lenalidomide as new therapeutics for the treatment of chronic lymphocytic leukemia. Leuk Lymphoma 51:27–38

D. Ribatti, *Angiogenesis and Anti-Angiogenesis in Hematological Malignancies*,
DOI 10.1007/978-94-017-8035-3, © Springer Science+Business Media Dordrecht 2014

Baban D, Murray J, Ear LH et al (1996) Quantitative analysis of vascular endothelial growth factor expression in chronic lymphocytic leukemia. Int J Oncol 8:29–34

Bairey O, Zimra Y, Kaganovsky E et al (2000) Microvessel density in chemosensitive and chemoresistant diffuse large B-cell lymphomas. Med Oncol 17:314–318

Bairey O, Boycov O, Kaganovsky E et al (2004) All three receptors fro vascular endothelial growth factor (VEGF) are expressed on B-chronic lymphocytic leukemia (CLL) cells. Leukemia Res 28:243–248

Barbarroja N, Torres LA, Luque MJ et al (2009) Additive effect of PTK787/ZK 222584, a potent inhibitor of VEGFR phosphorylation, with idarubicin in the treatment of acute myeloid leukemia. Exp Hematol 37:679–691

Barillé S, Akhoundi C, Collette M et al (1997) Metalloproteinases in multiple myeloma: production of matrix metalloproteinase-9 (MMP-9), activation of proMMP-2, and induction of MMP-1 by myeloma cells. Blood 90:1649–1655

Barlogie B, Desikan R, Eddlemon P et al (2001) Extended survival in advanced and refractory multiple myeloma after single-agent thalidomide: identification of prognostic factor in a phase-2 study of 169 patients. Blood 98:492–494

Basile A, Moschetta M, Di Tonno P et al (2013) Pentraxin 3 (PTX3) inhibits plasma cell/stromal cell cross-talk in the bone marrow of multiple myeloma patients. J Pathol 229:87–98

Baumann P, Muller K, Mandl-Weber S et al (2009) The peptide-semicarbazone S-2209, a representative of a new class of proteasome inhibitors, induces apoptosis and cell growth arrest in multiple myeloma cells. Br J Haematol 144(6):875–886

Bauvois J, Dumont J, Mathiot C et al (2002) Production of matrix metalloproteinase-9 in early B-CLL: suppression by interferons. Leukemia 16:791-798

Becker J, Covelo-Fernandez A, von Bonin F et al (2012) Specific tumor-stroma interactions of EBV-positive Burkitt's lymphoma cells in the chick chorioallantoic membrane. Vasc Cell 4:3

Belch A, Kouroukis CT, Crump M et al (2007) A phase II study of bortezomib in mantle cell lymphoma: the National Cancer Institute of Canada Clinical Trial Group trial IND.150. Ann Oncol 18:116–121

Bellamy WT (2001) Expression of vascular endothelial growth factor and its receptors in multiple myeloma and other hematopoietic malignancies. Semin Oncol 28:551–559

Bellamy WT, Richter L, Frutiger Y et al (1999) Expression of vascular endothelial growth factor and its receptors in hematopoietic malignancies. Cancer Res 59:728–733

Bellamy WT, Richter L, Sirjani D et al (2001) Vascular endothelial growth factor (VEGF) is an autocrine promoter of abnormal localized immature precursors (ALIP) and leukemia progenitor formation in myelodysplastic syndromes. Blood 97:1427–1434

Benezra M, Phillips E, Tilki D et al (2012) Serial monitoring of human systemic and xenograft models of leukemia using a novel vascular disrupting agents. Leukemia 26:1771–1778

Berardi S, Caivano A, Ria R et al (2012) Four proteins governing overangiogenic endothelial cell phenotype in patients with multiple myeloma are plausible therapeutic targets. Oncogene 31:2258–2269

Bergers G, Song S, Meyer-Morse N et al (2003) Benefits of targeting both pericytes and endothelial cells in the tumor vasculature with kinase inhibitors. J Clin Invest 111:1287–1295

Bertolini F, Fugetti L, Mancuso P et al (2000) Endostatin, an antiangiogenic drug, induces tumor stabilization after chemotherapy or anti-CD20 therapy in a NOD/SCID mouse model of human high-grade non-Hodgkin's lymphoma. Blood 96:282–287

Billadeau D, Ahmann G, Greipp P et al (1993) The bone marrow of multiple myeloma patients contains B cell populations at different stages of differentiation that are clonally related to malignant plasma cell. J Exp Med 178:1023–1031

Bisping G, Leo R, Wenning D et al (2003) Paracrine interactions of basic fibroblast growth factor and interleukin-6 in multiple myeloma. Blood 101:2775–2783

Bono P, Teerenhovi L, Joensuu H (2003) Elevated serum endostatin is associated with poor outcome in patients with non-Hodgkin lymphoma. Cancer 97:2762–2775

Børset M, Lien E, Espevik T et al (1996) Concomitant expression of hepatocyte growth factor/scatter factor and the receptor c-MET in human myeloma cell lines. J Biol Chem 271:24655–24661

Brandvold KA, Neiman P, Ruddell A (2000) Angiogenesis is an early event in the generation of myc-induced lymphomas. Oncogene 19:2780–2785

Brekken RA, Overholser JP, Stastny VA et al (2000) Selective inhibition of vascular endothelial growth factor (VEGF) receptor-2 (KDR/Flk-1) activity by a monoclonal anti-VEGF antibody blocks tumor growth in mice. Cancer Res 60:5117–5124

Broggini M, Marchini SV, Galliera E et al (2003) Aplidine, a new anticancer agent of marine origin, inhibits vascular endothelial growth factor (VEGF) secretion and blocks VEGF-VEGFR-1 (flt-1) autocrine loop in human leukemia cells MOLT-4A. Leukemia 17:52–59

Bruserud ØR, Olsnes AM et al (2007) Subclassification of patients with acute myelogeneous leukemia based on chemokine responsiveness and constitutive chemokine release by their leukemic cells. Haematologica 92:332–341

Burger JA, Ghia P, Rosenwald A et al (2009) The microenvironment in mature B-cell malignancies: a target for new treatment strategies. Blood 114:3367–3375

Cabebe E, Wakelee H (2006) Sunitinib: a newly approved small-molecule inhibitor of angiogenesis. Drugs Today (Barc) 42:387–398

Canale S, Cocco C, Frasson C et al (2011) Interleukin-27 inhibits pediatric B-acute lymphoblastic leukemia cell spreading in a preclinical model. Leukemia 25:1815–1824

Canioni D, Salles G, Mounier N et al (2008) High numbers of tumor-associated macrophages have an adverse prognostic value that can be circumvented by rituximab in patients with follicular lymphoma enrolled onto the GELA-GOELAMS FL-2000 trial. J Clin Oncol 26:440–446

Capillo M, Mancuso P, Gobbi A et al (2003) Continuous infusion of endostatin inhibits differentiation, mobilization and clonogenic potential of endothelial cell progenitors. Clin Cancer Res 9:377–382

Casanovas O, Hicklin DJ, Bergers G et al (2005) Drug resistance by evasion of antiangiogenic targeting of VEGF signaling in late-stage pancreatic islet tumors. Cancer Cell 8:299–309

Catlett-Falcone R, Landowski TH, Oshiro MM et al (1999) Constitutive activation of Stat3 signaling confers resistance to apoptosis in human U266 myeloma cells. Immunity 10:105–115

Chanan-Khan A, Miller KC, Takeshita K et al (2005) Results of phase 1 clinical trial of thalidomide in combination with fudarabine as initial therapy for patients with treatment-requiring chronic lymphocytic leukemia (CLL). Blood 106:3348–3352

Chanan-Khan A, Miller KC, Musial L et al (2006) Clinical efficacy of lenalidomide in patients with relapsed or refractory chronic lymphocytic leukemia: results of a phase II study. J Clin Oncol 24:5343–5349

Chauhan D, Uchiyama H, Akbarali Y et al (1996) Multiple myeloma cell adhesion-induced interleukin-6 expression in bone marrow stromal cells involves activation of NF-kappa B. Blood 87:1104–1112

Chauhan D, Pandey P, Ogata A et al (1997a) Dexamethasone induces apoptosis of multiple myeloma cells in a JNK/SAP kinase independent mechanism. Oncogene 15:837–843

Chauhan D, Kharbanda S, Ogata A et al (1997b) Interleukin-6 inhibits Fas-induced apoptosis and stress-activated protein kinase activation in multiple myeloma cells. Blood 89:227–234

Chen H, Treeweeke AT, West DC et al (2000) In vitro and in vivo production of vascular endothelial growth factor in chronic lymphocytic leukemia cells. Blood 96:3181–3187

Chen HX, Gore-Langton RE, Cheson BD (2001) Clinical trials referral resource: Current clinical trials of the anti-VEGF monoclonal antibody bevacizumab. Oncology 15:1023–1026

Cheriyath V, Hussein MA (2005) Osteopontin, angiogenesis and multiple myeloma. Leukemia 19:2203–2205

Clezardin P (2002) The antitumor potential of bisphosphonates. Semin Oncol 29(6 Suppl 21):33–42

Cocco C, Giuliani N, Di Carlo E et al (2010a) Interleukin-27 acts as multifunctional antitumor agent in multiple myeloma. Clin Cancer Res 16:4188–4197

Cocco C, Canale S, Frasson C et al (2010b) Interleukin-23 acts as antitumor agent on chicdhood B-acute lymphoblastic leukemia cells. Blood 116:3887–3898

Cocco C, di Carlo E, Zupo S et al (2012) Complementary IL-23 and IL-27 anti-tumor activities cause strong inhibition of human follicular and diffuse large B-cell lymphoma growth in vivo. Leukemia 26 1365–1374

Cohen T, Nahari D, Cerem LW et al (1996) Interleukin 6 induces the expression of vascular endothelial growth factor. J Biol Chem 271:736–741

Coleman M, Leonard J, Lyons L et al (2003) Treatment of Waldenstrom's macroglobulinemia with clarithromycin, low-dose thalidomide, and dexamethasone. Semin Oncol 30:270–274

Colla S, Morandi F, Lazzaretti M et al (2005) Human myeloma cells express the bone regulating gene Runx2/Cbfa1 and produce osteopontin that is involved in angiogenesis in multiple myeloma. Leukemia 19:2166–2176

Coluccia AM, Cirulli T, Neri P, Mangieri D et al (2008) Validation of PDGFR beta and c-Src tyrosine kinases as tumor/vessel targets in patients with multiple myeloma: preclinical efficacy of the novel, orally available inhibitor dasatinib. Blood 112:1346–1356

Corral LG, Haslett PA, Muller GW et al (1999) Differential cytokine modulation and T cell activation by two distinct classes of thalidomide analogues that are potent inhibitors of TNF-alpha. J Immunol 163:380–386

Crivellato E, Nico B, Vacca A et al (2002) Mast cell heterogeneity in B-cell non-Hodgkin's lymphomas: an ultrastructural study. Leuk Lymphoma 43:2201–2205

Crivellato E, Nico B, Vacca A et al (2003a) Ultrastructural analysis of mast cell recovery after secretion by piecemeal degranulation in B-cell non-Hodgkin's lymphoma. Leuk Lymphoma 44:517–521

Crivellato E, Nico B, Vacca A et al (2003b) B-cell non Hodgkin's lymphomas express heterogeneous patterns of neovascularization. Haematologica 88:671–678

Crump M, Coiffier B, Jacobsen ED et al (2008) Phase II trial of oral vorinostat (suberoylanilide hydroxamic acid) in relapsed diffuse large-B-cell lymphoma. Ann Oncol 19:964–969

Czuczman MS, Vose JM, Witzig TE et al (2011) The differential effect of lenalidomide monotherapy in patients with relapsed or refractory transformed non-Hodgkin lymphoma of distinct histological origin. Br J Haematol 154:477–481

Dal Lago L, D'Hondt V, Awada A (2008) Selected combination therapy with sorafenib: a review of clinical data and perspectives in advanced solid tumors. Oncologist 13:845–858

D'Amato RJ, Loughan MS, Flynn E et al (1994) Thalidomide is an inhibitor of angiogenesis. Proc Natl Acad Sci USA 91:4082–4085

Damiano JS, Cress AE, Hazlehurst LA et al (1999) Cell adhesion mediates drug resistance (CAM-DR): role of integrins and resistance to apoptosis in human myeloma cell line. Blood 93:1658–1667

Dankbar B, Padro T, Leo R et al (2000) Vascular endothelial growth factor and interleukin-6 in paracrine tumor-stromal cell interactions in multiple myeloma. Blood 95:2630–2636

Davidson JW, Hobbs BB, Fletch A (1973) The microcirculation unit of the mammalian lymph node. Bibl Anat 11:423–427

Davies FE, Raje N, Hideshima T et al (2001) Thalidomide and immunomodulatory derivatives augment natural killer cytotoxicity in multiple myeloma. Blood 98:210–216

Deaoanne CF, Bonjeean K, Servotte S et al (2002) Histone deacetylase inhibitors as anti-angiogenic agents altering vascular endothelial growth factor signaling. Oncogene 21:427–436

De Bont ES,F, Meeuwsen T et al (2002) Vascular endothelial growth factor secretion is an independent prognostic factor for relapse-free survival in pediatric acute myeloid leukemia patients. Clin Cancer Res 8:2856–2861

De Luisi AF, Coluccia AM et al (2011) Lenalidomide restrains motility and overangiogenic potential of bone marrow endothelial cells in patients with active multiple myeloma. Clin Cancer Res 17:1935–1946

De Palma MV, Roca C et al (2003) Targeting exogenous genes to tumor angiogenesis by transplantation of genetically modified hematopoietic stem cells. Nat Med 9:785–795

Derksen PW, de Gorter DJ, Meijer HP et al (2003) The hepatocyte growth factor/Met pathway controls proliferation and apoptosis in multiple myeloma. Leukemia 17:764–774

Desikan R, Li Z, Jagannath S (2002) Waldenstrom's macroglobulinemia: current therapy and future approaches. BioDrugs 16:201–207

de Vos SG, Dakhil SR et al (2009) Multicenter randomized phase II study of weekly or twice-weekly bortezomib plus rituximab in patients with relapsed or refractory follicular or marginal zone B-cell lymphoma. J Clin Oncol 27:5023–5030

Dias S, Hattori K, Zhu Z et al (2000) Autocrine stimulation of VEGFR-2 activates human leukemic cell growth and migration. J Clin Invest 106:511–521
Dimopoulos MA, Zervas K, Kouvatseas G et al (2001a) Thalidomide and dexamethasone combination for refractory multiple myeloma. Ann Oncol 12:991–995
Dimopoulos MA, Zomas A, Viniou NA et al (2001b) Treatment of Waldenstrom's macroglobulinemia with thalidomide. J Clin Oncol 19:3596–3601
Dimopoulos MA, Anagnostopouolos A, Weber D (2003) Treatment of plasma cell dyscrasias with thalidomide and its derivatives. J Clin Oncol 21:4444–4454
Ding W, Knox TR, Tschumper RC et al (2010) Platelet-derived growth factor (PDGF)-PDGF receptor interaction activates bone marrow-derived mesenchymal stromal cells derived from chronic lymphocytic leukemia: implications for an angiogenic switch. Blood 116:2984–2993
Di Raimondo FD, Azzaro MP, Palombo GA et al (2000) Angiogenic factors in multiple myeloma: higher levels in bone marrow than in peripheral blood. Haematologica 85:800–805
Dispenzieri A, Gertz MA, Lacy MQ et al (2006) A phase II trial of imatinb in patients with refractory/relapsed myeloma. Leuk Lymphoma 47:39–42
Djonov V, Schmid M, Tschanz SA et al (2000) Intussusceptive angiogenesis: its role in embryonic vascular network formation. Circ Res 86:286–292
Dogan A, Ngu LSP, NG SH et al (2005) Pathology and clinical features of angiommunoblastic T-cell lymphoma after successful treatment with thalidomide. Leukemia 19:873–875
Dominici M, Campioni D, Lanza F et al (2001) Angiogenesis in multiple myeloma: correlation between in vitro endothelial colonies growth (CFU-En) and clinical-biological features. Leukemia 15:171–176
Doussis-AnagnostopoulouI A, Talks KL, Turley H et al (2002) Vascular endothelial growth factor (VEGF) is expressed by neoplastic Hodgkin-Reed-Sternberg cells in Hodgkin's disease. J Pathol 197:677–683
Drewinko B, Alexanian R, Boyer H et al (1981) The growth fraction of human myeloma cells. Blood 57:333–338
Durie BG, Salmon SE (1975) High speed scintillation autoradiography. Science 190:1093–1095
Durie BG, Salmon SE, Moon TE (1980) Pretreatment tumor mass, cell kinetics. and prognosis in multiple myeloma. Blood 55:364–372
Duvic M, Talpur R, Ni X et al (2007) Phase 2 trial or oral vorinostat (suberoylanilide hydroxamic acid, SAHA) for refractory cutaneous T-cell lymphoma (CTCL). Blood 109:31–39
Ebos JM, Tran J, Master Z et al (2002) Imatinib mesylate (STI-571) reduces Bcr-Abl-mediated vascular endothelial growth factor secretion in chronic myelogenous leukemia. Mol Cancer Res 1:89–95
Edelman J, Klein-Hitpass L, Carpinteiro A et al (2008) Bone marrow fibroblasts induce expression of PI3K/NF-kappa B pathway genes and pro-angiogenic phenotype in CLL cells. Leuk Res 32:1565–1572
Ellis L, Pan Y, Smyth GK et al (2008) Histone deacetylase inhibitor panobinostat induces clinical responses with associated alterations in gene expression profiles in cutaneous T cell lymphoma. Clin Cancer Res 14:4500–4510
Fabbro D, Ruetz S, Bodis S et al (2000) PKC412-a protein kinase inhibitor with a broad therapeutic potential. Anticancer Drug Des 15:17–28
Fanelli M, Sarmiento R, Gattuso D et al (2003) Thalidomide: a new anticancer drug? Expet Opin Invest Drugs 12:1201–1215
Farinha P, Masoudi H, Skinnider BF et al (2005) Analysis of multiple biomarkers shows that lymphoma-associated macrophage (LAM) content is an independent predictor of survival in follcular lymphoma. Blood 106:2169–2174
Farohani M, Treweeke AT, Toh CH et al (2005) Autocrine VEGF mediates the anti-apoptotic effect of CD154 on CLL cells. Leukemia 19:524–530
Ferrajoli A, Lee BN, Schlette EJ et al (2008) Lenalidomide induces complete and partial remissions in patients with relapsed and refractory chronic lymphocytic leukemia. Blood 111:5291–5297
Ferretti E, Di Carlo E, Cocco C et al (2010) Direct inhibition of human acute myeloid leukemia cell growth by IL-12. Immolol Letters 133:99–105

Fiedler W, Graeven V, Ergun S et al (1997) Vascular endothelial growth factor, a possible paracrine growth factor in human acute myeloid leukemia. Blood 89:1870–1875
Fielder W, Mester R, Tinnefeld H et al (2003) A phase 2 clinical study of SU 5416 in patients with refractory acute myeloid leukemia. Blood 102:2763–2767
Fiedler W, Serve H, Dohner H et al (2005) A phase I study of SU11248 in the treatment of patients with refractory or resistant acute myeloid leukemia (AML) or not amendable to conventional therapy for the disease. Blood 105:986–993
Fiedler W, Mesters R, Heuser M et al (2010) An open-label, phase I study of cediranib (RECENTIN) in patients with acute myeloid leukemia. Leukemia 34:196–202
Fisher RI, Miller TP, Grogan TM (1998) New REAL clinical entities. Cancer J Sci Am 4(suppl 2):S5–S12
Fisher RI, Bernstein SH, Kahl BS et al (2006) Multicenter phase II study of bortezomib in patients with relapsed or refractory mantle cell lymphoma. J Clin Oncol 24:4867–4874
Fischer T, Stone RM, Deangelo DJ et al (2010) Phase IIB trial of oral Midostaurin (PKC412), the FMS-like tyrosine kinase 3 receptor (FLT3) and multi-targeted kinase inhibitor, in patients with acute myeloid leukemia and high-risk myelodysplastic syndrome with either wild-type or mutated FLT3. J Clin Oncol 28:4339–4345
Folkman J (1971) Tumor angiogenesis: therapeutic implications. N Engl J Med 285:1182–1186
Folkman J (1994) Angiogenesis and breast cancer. J Clin Oncol 12:441–443
Folkman J (2006) Antiangiogenesis in cancer therapy. Endostatin and its mechanism of action. Exp Cell Res 312:594–607
Folkman J, Grenspan HP (1975) Influence of geometry on control of cell growth. Biochem Biophys Acta 417:211–236
Folkman J, Watson K, Ingber D et al (1989) Induction of angiogenesis during the transition from hyperplasia to neoplasia. Nature 339:58–61
Fong TA, Shawver LK, Sun L et al (1999) SSU5416 is a potent and selective inhibitor of the vascular endothelial growth factor (Flk-1/KDR) that inhibits tyrosine kinase catalysis, tumor vascularization, and growth of multiple tumor types. Cancer Res 59:99–106
Foss B, Mentzoni L, Bruserud O (2001) Effects of vascular endothelial growth factor on acute myelogenous leukemia blasts. J Hematother Stem Cell Res 10:81–93
Foss HD, Araujo I, Demel G et al (1997) Expression of vascular endothelial growth factor in lymphomas and Castelman's disease. J Pathol 183:44–50
Ganjoo KN, An CS, Robertson MJ et al (2006) Rituximab, bevacizumab, and CHOP (RA-CHOP) in untreated diffuse large B-cell lymphoma: safety, biomarker and pharmacokinetics analysis. Leuk Lymphoma 47:998–1005
Garcia-Sanz R, Gonzales-Fraile MI, Sierra M et al (2002) The combination of thalidomide, cyclophosphamide and dexamethasone (ThaCyDex) is feasable and can be an option for relapsed/refractory multiple myeloma. Hematol J 3:43–48
Geitz H, Handt S, Zwingenberger K (1996) Thalidomide selectively modulates the density of cell surface molecules involved in the adhesion cascade. Immunopharmacology 31:213–221
Ghost AK, Secreto CR, Knox TR et al (2010) Circulating microvescicles in B-cell chronic lymphocytic leukemia can stimulate marrow stromal cells: implications for disease progression. Blood 115:1755–1764
Giles FJ, Bellamy WT, Estrov Z et al (2006) The anti-angiogenesis agent, AG-013736, has minimal activity in elderly patients with poor prognosis acute myeloid leukemia (AML) or myelodysplastic syndrome (MDS). Leuk Res 30:801–811
Gimbrone MA Jr, Leapman S, Cotran RS et al (1972) Tumor dormancy in vivo by prevention of neovascularization. J Exp Med 136:261–267
Giuliani N, Airoldi I (2011) Novel insights into the role of interleukin- 27 and interleukin-23 in human malignant and normal plasma cells. Clin Cancer Res 17(22):6963–6970
Giuliani N, Colla S, Lazzaretti M et al (2003) Proangiogenic properties of human myeloma cells: production of angiopoietin-1 and its potential relationship to myeloma-induced angiogenesis. Blood 102:638–645

Glade Bender JC, Kandel JJ et al (2004) Vascular remodeling and clinical resistance to antiangiogenic cancer therapy. Drug Resist Updat 7:289–300

Gora-Tybor J, Blonski JZ, Robak T (2005) Circulating vascular endothelial growth factor (VEGF) and its soluble receptors in patients with chronic lymphocytic leukemia. Eur Cytokine Netw 16:41–46

Goy A, Younes A, McLaughlin P et al (2005) Phase II study of proteasome inhibitor bortezomib in relapsed or refractory B-cell non-Hodgkin's lymphoma. J Clin Oncol 23:667–675

Grand EK, Chase AJ, Heath C et al (2004) Targeting FGFR3 in multiple myeloma: inhibition of t(4;14)-positive cells by SU5402 and PD173074. Leukemia 18:962–966

Gratzinger D, Zhao S, Marinelli RJ et al (2007) Microvessel density and expression of vascular endothelial growth factor and its receptors in diffuse large B-cell lymphoma subtypes. Am J Pathol 170: 1362–1369

Gratzinger D, Zhao S, Tibshirani RJ et al (2008) Prognostic significance of VEGF, VEGF receptors, and microvessel density in diffuse large B cell lymphoma treated with antracycline-based chemotherapy. Lab Invest 88:38–47

Gunsilius E, Duba HC, Petzer AL et al (2000) Evidence from a leukaemia model for maintenance of vascular endothelium by bone-marrow-derived endothelial cells. Lancet 355:1688–1691

Habermann TM, Lossos IS, Justice G et al (2009) Lenalidomide oral monotherapy produces a high response rate in patients with relapsed or refractory mantle cell lymphoma. Br J Haematol 145:344–349

Hallek M, Bergsagel PL, Anderson KC (1998) Multiple myeloma: increasing evidence for a multistep transformation process. Blood 91:3–21

Hanahan D, Folkman J (1996) Patterns and emerging mechanisms of the angiogenic switch during tumorigenesis. Cell 86:353–364

Hatfiled KJ, Hovland R, Øyan AM et al (2008) Release of angiopoietin-1 by primary human acute myelogenous leukemia cells is associated with mutations of nucleophosmin, increased by bone marrow stromal cells and possibly antagonized by high systemic angiopoietin-2 levels. Leukemia 22:287–293

Hatfield K, Oyan AM, Ersvaer E et al (2009) Primary human acute myeloid leukaemia cells increase the proliferation of microvascular endothelial cells through the release of soluble mediators. Br J Haematol 144:53–68

Hatfield KJ, Bedringsaas SL, Ryningen A et al (2010a) Hypoxia increases HIF-1α expression and constitutive cytokine release by primary human acute myeloid leukaemia cells. Eur Cytokine Netw 21:154–164

Hatfield KJ, Reikvam H, Bruserud Ø (2010b) The crosstalk between the matrix metalloprotease system and the chemokine network in acute myeloid leukemia. Curr Med Chem 7:4448–4461

Hatfield KJ, Evensen L, Reikvam H et al (2012) Soluble mediators released by acute myeloid leukemia cells increase capillary-like networks. Eur J Haematol 89:478–490

Hatfill SJ, Fester ED, de Beer DP et al (1991) Induction of morphological differentiation of the human leukemic cell line K562 by exposure to thalidomide metabolities. Leukemia Res 15:129–136

Hayashibara T, Yamada Y, Onimaru Y et al (2002) Matrix metalloproteinase-9 and vascular endothelial growth factor: a possible link in adult T-cell leukemia cell invasion. Br J Haematol 116:94–102

Hazar B, Paydas Z, Zorludemir S et al (2003) Prognostic significance of microvessel density and vascular endothelial growth factor (VEGF) expression in non Hodgkin's lymphoma. Leuk Lymphoma 44:2089–2093

Hedvat CV, Comenzo RL, Teruya-Feldstein J et al (2003) Insights into extramedullary tumour cell growth revealed by expression profiling of human plasmacytomas and multiple myeloma. Br J Haematol 122:728–744

Heider U, Kaiser M, Sterz J et al (2006) Histone deacetylase inhibitors reduce VEGF production and induce growth suppression and apoptosis in human mantle cell lymphoma. Eur J Haematol 76:42–50

Henk HJ, Teitelbaum A, Perez JR et al (2012) Persistency with zoledronic acid is associated with clinical benefit in patients with multiple myeloma. Am J Hematol 87(5):490–495

Herman PG (1980) Microcirculation of organized lymphoid tissue. Monogr Allergy 16:126–142

Hideshima T, Chauhan D, Shima Y et al (2000) Thalidomide and its analogs overcome drug resistance of human multiple myeloma cells to conventional therapy. Blood 96:2943–2950

Hideshima T, Nakamura N, Chauhan D et al (2001a) Biologic sequelae of interleukin-6 induced PI3-K/Akt signaling in multiple myeloma. Oncogene 20:5991–6000

Hideshima T, Chauhan D, Schlossman R et al (2001b) The role of tumor necrosis factor alpha in the pathophysiology of human multiple myeloma: therapeutic applications. Oncogene 20:4519–4527

Hideshima T, Chauhan D, Richardson P et al (2002a) NF-kappa B as a therapeutic target in multiple myeloma. J Biol Chem 277:16639–16647

Hideshima T, Chauhan D, Hayashi T et al (2002b) The biological sequelae of stromal cell-derived factor-1alpha in multiple myeloma. Mol Cancer Ther 1:539–544

Hideshima T, Mitsiades C, Akiyama M et al (2003) a) Molecular mechanisms mediating antimyeloma activity of proteasome inhibitor PS-341. Blood 101:1530–1534

Hideshima T, Chauhan D, Hayashi T et al (2003) b) Proteasome inhibitor PS-341 abrogates IL-6 triggered signaling cascades via caspase-dependent downregulation of gp130 in multiple myeloma. Oncogene 22:8386–8393

Hideshima T, Podar K, Chauhan D et al (2005) Cytokines and signal transduction. Best Pract Res Clin Haematol 18:509–524

Ho CL, Sheu LF, Li CY (2002) Immunohistochemical expression of basic fibroblast growth factor, vascular endothelial growth factor, and their receptors in stage IV non-Hodgkin lymphoma. Appl Immunohistochem Mol Morphol 10:316–321

Holash J, Davis S, Papadopoulos N et al (2002) VEGF-Trap: A VEGF blocker with potent antitumor effects. Proc Natl Acad Sci USA 99:11393–11398

Hose D, Moraaux J, Meissner T et al (2009) Induction of angiogenesis by normal and malignat plasma cells. Blood 114:128–143

Hu L, Shi Y, Hsu JH et al (2003) Downstream effectors of oncogenic ras in multiple myeloma cells. Blood 101:3126–3135

Huang X, Bai X, Cao Y et al (2010) Lymphoma endothelium preferentially expresses Tim-3 and facilitates the progression of lymphoma by mediating immune evasion. J Exp Med 207:505–520

Hu-Lowe DD, Zou HY, Grazzini ML et al (2008) Nonclinical antiangiogenesis and antitumor activities of abitini (AG-013736), an oral, potent, and selective inhibitor of vascular endothelial growth factor receptor tyrosine kinases 1,2,3. Clin Cancer Res 14:7272–7283

Hussong JW, Rodgers GM, Shami PJ (2000) Evidence of increased angiogenesis in patients with acute myeloid leukemia. Blood 95:309–313

Igreja C, Courinha M, Cachaco AS et al (2007) Characterization and clinical relevance of circulating and biopsy-derived endothelial progenitor cells in lymphoma patients. Haematologica 92:469–477

Imai K, Takaoka A (2006) Comparing antibody and small-molecule therapies for cancer. Nat Rev Cancer 6:714–727

Inai T, Mancuso M, Hashizume H et al (2004) Inhibition of vascular endothelial growth factor (VEGF) signaling in cancer causes loss of endothelial fenestrations, regression of tumor vessels, and appearance of basement membrane ghosts. Am J Pathol 165:35–52

Ingber DE, Folkman J (1989) How does extracellular matrix control capillary morphogenesis? Cell 58:803–805

Iwasaki T, Hamano T, Ogata A et al (2002) Clinical significance of vascular endothelial growth factor and hepatocyte growth factor in multiple myeloma. Br J Haematol 116:796–802

Jagannath S, Barlogie B, Berenson J et al (2004) A phase 2 study of two doses of bortezomib in relapsed or refractory myeloma. Br J Haematol 127:165–172

Jia HY, Wu JX, Zhu XF et al (2009) ZD6474 inhibits Src kinase leading to apoptosis of imatinib-resistant K562 cells. Leuk Res 33:1512–1519

Jørgensen JM, Sørensen FB, Bendix K et al (2007) Angiogenesis in non Hodgkin's lymphoma: clinico-pathological correlations and prognostic significance in specific subtypes. Leuk Lymphoma 48:584–595

Jourdan M, Veyrune JL, Vos JD et al (2003) A major role for Mcl-1 antiapoptotic protein in the IL-6-induced survival of human myeloma cells. Oncogene 22:2950–2959

Jung YD, Mansfield PF, Akagi M et al (2002) Effects of combination anti-vascular endothelial growth factor receptor and anti-epidermal growth factor receptor therapies on the growth of gastric cancer in a nude mouse model. Eur J Cancer 38:1133–1140

Kamiguti AS, Lee ES, Till KJ et al (2004) The role of matrix metalloproteinase 9 in the pathogenesis of chronic lymphocytic leukaemia. Br J Haematol 125:128–140

Kane RC, Bross PF, Farrell AT et al (2003) Velcade: USFDA approval for the treatment of multiple myeloma progressing on prior therapy. Oncologist 8:508–513

Kane RC, Farrell AT, Saber H et al (2006a) Sorafenib for the treatment of advanced renal cell carcinoma. Clin Cancer Res 12:7271–7278

Kane RC, Farrell AT, Sridhara R et al (2006b) United States Food and Drug Administration approval summary: bortezomib for the treatment of progressive multiple myeloma after one prior therapy. Clin Cancer Res 12(10):2955–2960

Karashima T, Inoue K, Fukata S et al (2007) Blockade of the vascular endothelial growth factor-receptor 2 pathway inhibits the growth of human renal cell carcinoma, RBM1-IT4, in the kidney but not in the bone of nude mice. Int J Oncol 30:937–945

Karp JE, Gojo I, Pili R et al (2004) Targeting vascular endothelial growth factor for relapsed and refractory adult acute myelogeneous leukemia: therapy with sequential 1-β-D-arabinofuranosylcytosine, mitoxantrone, and bevacizumab. Clin Cancer Res 10:3577–3585

Kaufmann H, Raderer M, Wöhrer S et al (2004) Antitumor activity of rituximab plus thalidomide in patients with relapsed/refractory mantle cell lymphoma. Blood 104:2269–2271

Kay NE, Bone ND, Tschumper RC et al (2002) B-CLL cells are capable of synthesis and secretion of both pro- and anti-angiogenic molecules. Leukemia 16:911–919

Keats JJ, Reiman T, Maxwell CA et al (2003) In multiple myeloma, t(4;14)(p16;q32) is an adverse prognostic factor irrespective of FGFR3 expression. Blood 101:1520–1529

Keifer JA, Guttridge DC, Ashburner BP et al (2001) Inhibition of NF-kB activity by thalidomide through suppression of IkB kinase activity. J Biol Chem 276:22382–22387

Kenyon BM, Browne F, D'Amato RJ (1997) Effects of thalidomide and related metabolites in a mouse corneal model of neovascularization. Exp Eye Res 64:971–978

Kerbel KS, Viloria-Petit A, Klement G et al (2000) 'Accidental' anti-angiogenic drugs: anti-oncogene directed signal transduction inhibitors and conventional chemotherapeutic agents as examples. Eur J Cancer 36:1248–1257

Kim ES, Serur A, Huang J et al (2002) Potent VEGF blockade causes regression of coopted vessels in a model of neuroblastoma. Proc Natl Acad Sci USA 99:11399–11404

Kim KJ, Li B, Winer J et al (1993) Inhibition of vascular endothelial growth factor induced angiogenesis suppresses tumor growth in vivo. Nature 362:841–844

Kini AR (2004) Angiogenesis in leukemia and lymphoma. Cancer Treat Res 121:221–238

Kini AR, Kay NE, Peterson LC (2000) Increased bone marrow angiogenesis in B-cell chronic lymphocytic leukemia. Leukemia 14(8):1414–1418

Klein B, Zhang XG, Lu ZY et al (1995) Interleukin-6 in human multiple myeloma. Blood 85:863–872

Knapper S, Burnett AK, Littlewoord T et al (2006) A phase 2 trial of the FLT3 inhibitor lestaurtinib (CEP701) as first-line treatment for older patients with acute myeloid leukemia not considered fit for intensive chemotherapy. Blood 108:3262–3270

König A, Menzel T, Lynen S et al (1997) Basic fibroblast growth factor (bFGF) upregulates the expression of bcl-2 in B cell chronic lymphocytic leukemia cell lines resulting in delaying apoptosis. Leukemia 11:258–265

Koomagi R, Zintl F, Sauerbrey A et al (2001) Vascular endothelial growth factor in newly diagnosed and recurrent childhood acute lymphoblastic leukemia as measured by real-time quantitative polymerase chain reaction. Clin Cancer Res 7:3381–3384

Korkolopoulou P, Aposstodilou E, Pavlopoullos PM et al (2001) Prognostic evaluation of the microvascular network in myelodyslastic syndromes. Leukemia 15:1369–1376

Koster A, Raemaekers JMM (2005) Angiogenesis in malignant lymphoma. Curr Opin Oncol 17:611–616

Koster A, van Krieken JH, Mackenzie MA et al (2005b) Increased vascularization predicts favorable outcome in follicular lymphoma. Clin Cancer Res 11:154–161

Kruizinga RC, de Jonge HJ, Kampen KR et al (2011) Vascular endothelial growth factor isoform mRNA expression in pediatric acute myeloid leukemia. Pediatr Blood Cancer 56:294–297

Kumar S, Witzig TE, Timm M et al (2003a) Expression of VEGF and its receptors by myeloma cells. Leukemia 17:2025–2031

Kumar S, Gertz MA, Dispenzieri A et al (2003b) Response rate, durability of response, and survival after thalidomide therapy for relapsed mutiple myeloma. Mayo Clinic Proc 78:34–39

Kumar S, Witzig TE, Timm M et al (2004) Bone marrow angiogenic ability and expression of angiogenic cytokines in myeloma: evidence favoring loss of marrow angiogenesis inhibitory activity with disease progression. Blood 104:1159–1165

Kuramoto K, Sakai A, Shigemasa K et al (2002) High expression of MCL1 gene related to vascular endothelial growth factor is associated with poor outcome in non-Hodgkin's lymphoma. Br J Haematol 116:158–161

Kuzu I, Beksac M, Arat M et al (2004) Bone marrow microvessels density (MVD) in adult acute myeloid leukemia (AML): therapy induced changes and effects on survival. Leuk Lymphoma 45:1185–1190

Lee YK, Bone ND, Strege AK et al (2004) VEGF receptor phosphorylation status and apoptosis in modulation by a green tea component, epigallocatechin-3-gallate (EGCG), in B-cell chronic lymphocytic leukemia. Blood 104:788–794

Lee YK, Shanafelts TD, Bone ND et al (2005) VEGF receptors on chronic lymphocytic leukemia (CLL) B cells interact with STAT1 and 3: implication for apoptosis resistance. Leukemia 19:513–523

Le Gouill SP, Amiot M et al (2004) VEGF induces Mcl-1 up-regulation and protects multiple myeloma cells against apoptosis. Blood 104:2886–2892

Leenders WP, Küsters B, Verrijp K et al (2004) Antiangiogenic therapy of cerebral melanoma metastases results in sustained tumor progression via vessel co-option. Clin Cancer Res 10:6222–6230

Levis M, Allebach J, Tse KF et al (2002) A FLT-3-targeted tyrosine kinase inhibitor is cytotoxic to leukemia cells in vitro and in vivo. Blood 99:3855–3891

Lichtenstein A, Tu Y, Fady C et al (1995) Interleukin-6 inhibits apoptosis of malignant plasma cells. Cell Immunol 162:248–255

Liersch R, Schliemann C, Bieker R et al (2008) Expression of VEGF-C and its receptor VEGFR-3 in the bone marrow of patients with acute myeloid leukaemia. Leuk Res 32:954–961

Liesveld JL, Rosell KE, Lu C et al (2005) Acute myelogenous leukemia-microenvironment interactions: role of endothelial cells and proteasome inhibition. Hematology 10:483–494

Lin B, Podar K, Gupta D et al (2002) The vascular endothelial growth factor receptor tyrosine kinase inhibitor PTK787/ZK222584 inhibits growth and migration of multiple myeloma cells in the bone marrow microenvironment. Cancer Res 62:5019–5026

Lin LI, Lin DT, Chang CJ et al (2002) Marrow matrix metalloproteinases (MMPs) and tissue inhibitors of MMP in acute leukaemia: potential role of MMP-9 as a surrogate marker to monitor leukaemic status in patients with acute myelogenous leukaemia. Br J Haematol 117:835–841

List A, Kurtin S, Roe DJ et al (2005) Efficacy of lenalinomide in myelodysplastic syndromes. N Engl J Med 352:549–557

Litwin C, Leong KG, Zapf R et al (2002) Role of microenvironment in promoting angiogenesis in acute leukemia. Am J Hematol 70:22–30

Liu H, Tamashiro S, Baritaki S et al (2012) TRAF6 activation in multiple myeloma: a potential therapeutic target. Clin Lymphoma Myeloma Leuk 12:155–163

Loges S, Tinnefeld H, Metzner A et al (2006) Downregulation of VEGF-A, STAT5 and AKT in acute myeloid leukemia blasts of patients treated with SU5416. Leuk Lymphoma 47:2601–2609

Lundberg LG, Lerner P, Sundelin R et al (2000) Bone marrow in polycythemia vera, chronic myelocytic leukemia and myelofibrosis has an increased vascularity. Am J Pathol 157:15–19

Lyden D, Hattori K, Dias S et al (2001) Impaired recruitment of bone-marrow-derived endothelial and hematopoietic precursor cells blocks tumor angiogenesis and growth. Nat Med 7:1194–1201

Madlambayan GJ, Meacham AM, Hosaka K et al (2010) Leukemia regression by vascular disruption and antiangiogenic therapy. Blood 116:1539–1547

Maffei R, Martinelli S, Castelli I et al (2010a) Increased angiogenesis induced by chronic lymphocytic leukemia B cells is mediated by leukemia-derived Ang-2 and VEGF. Leuk Res 34:312–321

Maffei R, Martinelli S, Santachiara R et al (2010b) Angiopoietin-2 plasma dosage predicts time to first treatment and overall survival in chronic lymphocytic leukemia. Blood 116:2984–2993

Majumbar S, Lamothe B, Aggarwal BB (2002) Thalidomide suppresses NF-kB activation induced by TNF-a and H_2O_2, but not that activated by ceramide, lipopolysaccharides, or phorbol ester. J Immunol 168:2644–2651

Mancuso P, Burlini A, Pruneri G et al (2001) Resting and activated endothelial cells are increased in the peripheral blood of cancer patients. Blood 97:3658–3661

Mangieri D, Nico B, Benagiano V et al (2008) Angiogenic activity of multiple myeloma endothelial cells in vivo in the chick embryo chorioallantoic membrane assay is associated to a down-regulation in the expression of endogenous endostatin. J Cell Mol Med 12:1023–1028

Marks MG, Shi J, Fry MO et al (2002) Effects of putative hydroxylated thalidomide metabolities on blood vessel density in the chorioallantoic membrane (CAM) assay and on tumor and endothelial cell proliferation. Biol Pharmacol Bull 25:597–604

Mayerhofer M, Valent P, Sperr WR et al (2002) BCR/ABL induces expression of vascular endothelial growth factor and its transcriptional activator, hypoxia inducible factor 1-alpha, through a pathway involving phosphoinostide 3-kinase and the mammalian target of rapamycin. Blood 100:3767–3775

Mazur G, Woźniak Z, Wróbel T et al (2004) Increased angiogenesis in cutaneous T-cell lymphomas. Pathol Oncol Res 10:34–36

Mc Hugh SM, Rifkin IR, Deighton J et al (1995) The immunosuppressive drug thalidomide induces T helper cell type 2 (Th2) and concomitantly inhibits Th1 cytokine production in mitogen- and antigen-stimulated human peripheral blood mononuclear cell cultures. Clin Exp Immunol 99:160–167

Medina V, Edmonds B, Young GP et al (1997) Induction of caspase-3 protease activity and apoptosis by butyrate and trichostatin A (inhibitors of histone deacetylase): dependence on protein synthesis and synergy with a mitochondrial/ctythochrome c-dependent pathway. Cancer Res 57:3697–3707

Melve GK, Ersvssr E, Kittang AO et al (2011) The chemokine system in allogenic stem-cell transplantation: a possible therapeutic target? Expert Rev Hematol 4:563–576

Mendel D, Laird AD, Xin X et al (2003) In vivo antitumor activity of SU11248, a novel tyrosine kinase inhibitor targeting vascular endothelial growth factor and platelet-derived growth factor receptors; determination of a pharmacokinetic/pharmacodynamic relationship. Clin Cancer Res 9:327–337

Menzel T, Rahman Z, Calleja E et al (1996) Elevated intracellular level of basic fibroblast growth factor correlates with stage of chronic lymphocytic leukemia and is associated with resistance to fludarabine. Blood 87:1056–1063

Mesters RM, Padro T, Bieker R et al (2001) Stable remission after administration of the receptor tyrosine kinase inhibitor SU5416 in patients with refractory acute myeloid leukemia. Blood 98:341–343

Mesters RM, Zahiragic L, Schliemannn C et al (2007) Bevacizumab reduces VEGF expression in patients with relapsed and refractory acute myeloid leukemia without clinical antileukemic activity. Leukemia 21:1310–1312

Metzelder SK, Wollmer E, Neubauer A et al (2010) Sorafenib in relapsed and refractory FLT-3-IDT positive acute myeloid leukemia: a novel treatment option. Deutsche Med Woch 135:1852–1856

Millauer B, Shawver LK, Plate KH et al (1994) Glioblastoma growth inhibited in vivo by a dominat-negative Flk-1 mutant. Nature 367:576–579

Mitsiades CS, Mitsiades N, Poulaki V et al (2002a) Activation of NF-kappaB and upregulation of intracellular anti-apoptotic proteins via the IGF-1/Akt signaling in human multiple myeloma cells: therapeutic implications. Oncogene 21:5673–5683

Mitsiades N, Mitsiades CS, Poulaki V et al (2002b) Apoptotic signaling induced by immunomodulatory thalidomide analogs in human multiple myeloma cells: therapeutic implications. Blood 99:4525–4530

Mitsiades CS, Mitsiades NS, McMullan CJ et al (2004) Inhibition of the insulin-like growth factor receptor-1 tyrosine kinase activity as a therapeutic strategy for multiple myeloma, other hematologic malignancies, and solid tumors. Cancer Cell 5:221–230

Molica S, Vitelli C, Levato D et al (1999) Increased serum levels of vascular endothelial growth factor predict risk of progression in early B-cell chronic lymphocytic leukemia. Br J Haematol 107:605–610

Molica S, Santoro R, Digiesi G et al (2000) Vascular endothelial growth factor isoforms 121 and 165 are expressed on B-chronic lymphocytic leukemia cells. Haematologica 85:1106–1108

Molica S, Vacca A, Ribatti D et al (2002) Prognostic value of enhanced bone marrow angiogenesis in early B-cell chronic lymphocytic leukemia. Blood 100:3344–3351

Molica S, Vitelli G, Levato D et al (2003a) Increased serum levels of matrix metalloproteinase-9 predict clinical outcome of patients with early B-cell chronic lymphocytic leukemia. Eur J Haematol 70:373–378

Molica S, Vacca A, Crivellato E et al (2003b) Tryptase positive mast cells predict clinical outcome of patients with early B-cell chronic lymphocytic leukemia. Eur J Haematol 71:137–139

Molica S, Montillo M, Ribatti D et al (2007a) Intense reversal of bone marrow angiogenesis after sequential fludarabine-induction and alemtuzumab-consolidation therapy in advanced chronic lymphocytic leukemia. Haematologica 92:1367–1374

Molica S, Cutrona G, Vitelli G et al (2007b) Markers of increase angiogenesis and their correlation with biological parameters identifying high risk patients in early B-cell chronic lymphocytic leukemia. Leuk Res 31:1575–1578

Moller DR, Wysocka M, Greenlee BM et al (1997) Inhibition of IL-12 production by thalidomide. J Immunol 159:5157–5161

Monestiroli S, Mancuso P, Burlini A et al (2001) Kinetics and viability of circulating endothelial cells as surrogate angiogenesis marker in an animal model of human lymphoma. Cancer Res 61:4341–4344

Morabito A, Piccirillo MC, Falasconi F et al (2009) Vandetanib (ZD6474), a dual inhibitor of vascular endothelial growth factor receptor (VEGFR) and epidermal growth factor receptor (EGFR) tyrosine kinases: current status and future directions. Oncologist 14:378–390

Moreira AL, Corral LG, Ye W et al (1997) Thalidomide and thalidomide analogs reduce HIV type 1 receptor replication in human macrophages in vitro. AIDS Res Hum Retrov 10:857–863

Morgan GJ, Davies FE, Gregory WM et al (2012) Effects of induction and maintenance plus long-term bisphosphonates on bone disease in patients with multiple myeloma: MRC Myeloma IX trial. Blood 119:5374–5383

Moschetta M, Di Pietro G, Ria R et al (2010) Bortezomib and zoledronic acid on angiogenic and vasculogenic activities of bone marrow macrophages in patients with multiple myeloma. Eur J Cancer 46(2):420–429

Motro B, Itin A, Sachs L et al (1990) Pattern of interleukin 6 gene expression in vivo suggests a role for this cytokine in angiogenesis. Proc Natl Acad Sci USA 87:3092–3096

Munshi NC, Hideshima T, Carrasco D et al (2004) Identification of genes modulated in multiple myeloma using genetically identical twin samples. Blood 103:1799–1806

Nakagawa M, Kaneda T, Arawaka T et al (2000) Vascular endothelial growth factor (VEGF) directly enhances osteoclastic bone resorption and survival of mature osteoclasts. FEBS Lett 473:161–164

Nakayama T, Yao L, Tosato G (2004) Mast cell-derived angiopoietin-1 plays a critical role in the growth of plasma cell tumors. J Clin Invest 114:1317–1325

Neufeld G, Cohen T, Gengrinovitch S et al (1999) Vascular endothelial growth factor (VEGF) and its receptors. FASEB J 13:9–22

Nico B, Mangieri D, Crivellato E et al (2008) Mast cells contribute to vasculogenic mimicry in multiple myeloma. Stem Cells Dev 17:19–22

Nico B, Annese T, Tamma R et al (2012) Aquaporin-4 expression in primary human central nervous system lymphomas correlates with tumour cell proliferation and phenotypic heterogeneity of the vessel wall. Eur J Cancer 48:772–781

Niemoller K, Jakob C, Heider V et al (2003) Bone marrow angiogenesis and its correlation with other disease characteristics in multiple myeloma in stage I versus stage II-III. J Cancer Res Clin Oncol 129:234–238

Niida S, Kaku M, Amano H et al (1999) Vascular endothelial growth factor can substitute for macrophage colony-stimulating factor in the support of osteoclastic bone resorption. J Exp Med 190:293–298

Niitsu N, Okamato M, Nakamine H et al (2002) Simultaneous elevation of the serum concentrations of vascular endothelial growth factor and interleukin-6 as independent predictors of prognosis in aggressive non-Hodgkin's lymphoma. Eur J Pharmacol 68:91–100

Nishioka C, Ikezoe T, Takeshita A et al (2007) ZD6474 induces growth arrest and apoptosis of human leukemia cells, which is enhanced by concomitant use of a novel MEK inhibitor, AZD6244. Leukemia 21:1308–1310

Norén-Nyström U, Heyman M, Frisk P et al (2009) Vascular density in childhood acute lymphoblastic leukaemia correlates to biological factors and outcome. Br J Haematol 146:521–530

O'Connor OA, Portlock C, Moskowitz C et al (2010) Time to treatment response in patients with follicular lymphoma treated with bortezomib is longer compared with other histologic subtypes. Clin Cancer Res 16:719–726

O'Farrell AM, Abrams TJ, Yuen HA et al (2003) SU11248 is a novel FLT3 tyrosine kinase inhibitor with potent activity in vitro and in vivo. Blood 101:3597–3605

Ogata A, Chauhan D, Teoh G et al (1997) IL-6 triggers cell growth via the Ras-dependent mitogen-activated protein kinase cascade. J Immunol 159:2212–2221

Padro T, Bieker R, Ruiz S et al (2002) Overexpression of vascular endothelial growth factor (VEGF) and its cellular receptor KDR (VEGFR-2) in the bone marrow of patients with acute myeloid leukemia. Leukemia 16:1302–1310

Paesler J, Gehrke I, Gandhirajan RK et al (2010) The vascular endothelial growth factor receptor tyrosine kinase inhibitors vatalanib and pazopanib potently induce apoptosis in chronic lymphocytic leukemia in vitro and in vivo. Clin Cancer Res 16:3390–3398

Paku S, Paweletz N (1991) First steps of tumor-related angiogenesis. Lab Invest 65:334–346

Palumbo A, Bertola A, Falco P et al (2004) Efficacy of low-dose thalidomide and dexamethasone as first salvage regimen in multiple myeloma. Hematol J 5:318–324

Panayiotidis P, Jones D, Ganeshaguru K et al (1996) Human bone marrow stromal cells prevent apoptosis and support the survival of chronic lymphocytic leukaemia cells in vitro. Br J Haematol 92:97–103

Pandiella A, Carvajal-Vergara X, Tabera S et al (2003) Imatinib mesylate (ST 1571) inhibits multiple myeloma cell proliferation and potentiates the effect of common antimyeloma agents. Br J Haematol 123:858–868

Partida-Sanchez S, Favila-Castillo L, Pedraza-Sanchez S et al (1998) IgG antibody subclasses, tumor necrosis factor and IFN-gamma levels in patients with type II lepra reaction on thalidomide treatment. Int Arch Allergy Immunol 116:60–66

Passalidou E, Stewart M, Trivella M et al (2003) Vascular patterns in reactive lymphoid tissue and in non-Hodgkin's lymphoma. Br J Cancer 88:553–559

Paterson JL, Li Z, Wen XY et al (2004) Preclinical studies of fibroblast growth factor receptor 3 as a therapeutic target in multiple myeloma. Br J Haematol 124:595–603

Pazgal I, Zimr Y, Tzabar C et al (2002) Expression of basic fibroblast growth factor is associated with poor outcome in non-Hodgkin's lymphoma. Br J Cancer 86:1770–1775

Pellegrino A, Ria R, Di Pietro G et al (2005) Bone marrow endothelial cells in multiple myeloma secrete CXC-chemokines that mediate interactions with plasma cells. Br J Haematol 129:248–256

Perez-Atayde AR, Sallan SE, Tedrow U et al (1997) Spectrum of tumor angiogenesis in the bone marrow of children with acute lymphocytic leukemia. Am J Pathol 150:815–821

Petit I, Karajannis MA, Vincent L et al (2008) The microtubule-targeting agent CA4P regresses leukemic xenografts by disrupting interaction with vascular cells and mitochondrial-dependent cell death. Blood 111:1951–1961

Piva R, Ruggeri B, Williams M et al (2008) CEP-18770: A novel, orally active proteasome inhibitor with a tumor-selective pharmacologic profile competitive with bortezomib. Blood 111(5):2765–2775

Podar K, Anderson KC (2005) The pathophysiologic role of VEGF in hematologic malignancies: therapeutic implications. Blood 105:1383–1395

Podar K, Tai YT, Davies FE et al (2001) Vascular endothelial growth factor triggers signaling cascades mediating multiple myeloma cell growth and migration. Blood 98:428–435

Podar K, Tai YT, Lin BK et al (2002) Vascular endothelial growth factor (VEGF)-induced migration of multiple myeloma cells is associated with b1-integrin- and PI3-kinase-dependent PKCa activation. J Biol Chem 277:7875–7881

Podar K, Catley LP, Tai YT et al (2004) GW654652, the pan-inhibitor of VEGF receptors, blocks the growth and migration of multiple myeloma cells in the bone marrow microenvironment. Blood 103:3474–3479

Podar K, Tonon G, Sattler M et al (2006) The small-molecule VEGF receptor inhibitor pazopanib (GW786034B) targets both tumor and endothelial cells in multiple myeloma. Proc Natl Acad Sci USA 103:19478–19483

Presta LG, Chen H, O'Connors SJ et al (1997) Humanization of an anti-vascular endothelial growth factor monoclonal antibody for the therapy of solid tumors and other disorders. Cancer Res 57:4593–4599

Prince HM, Hönemann D, Spencer A et al (2009) Vascular endothelial growth factor inhibition is not an effective therapeutic strategy for relapsed or refractory multiple myeloma: a phase 2 study of pazopanib (GW786034). Blood 113:4819–4820

Pro B, Younes A, Albitar M et al (2004) Thalidomide for patients with recurrent lymphoma. Cancer 100:1186–1189

Pui CH, Relling MV, Dowing JR (2004) Acute lymphoblastic leukemia. N Engl J Med 350:1535–1548

Pulè MA, Gullmann C, Dennis D et al (2002) Increased angiogenesis in bone marrow of children with acute lymphoblastic leukaemia has not prognostic signifcance. Br J Haematol 118:991–998

Pulkki K, Pelliniemi TT, Kajamaki A et al (1996) Soluble interleukin-6 receptor as a prognostic factor in multiple myeloma. Br J Haematol 92:370–374

Qiang YW, Kopantzev E, Rudikoff S (2002) Insulin-like growth factor-I signaling in multiple myeloma: downstream elements, functional correlates, and pathway cross-talk. Blood 99:4138–4146

Qing J, Du X, Chen Y et al (2009) Antibody-based targeting of FGFR3 in bladder carcinoma and t(4;14)-positive multiple myeloma in mice. J Clin Invest 119:1216–1229

Rabitsch W, Sperr WR, Lechner K et al (2004) Bone marrow microvessel density and its prognostic significance in AML. Leuk Lymphoma 45 :1369–1373

Raje N, Kumar S, Hideshima T et al (2004) Combination of the mTOR inhibitor rapamycin and CC-5013 has synergistic activity in multiple myeloma. Blood 104:4188–4193

Rajkumar SV, Sonneveld P (2009) Front-line treatment in younger patients with multiple myeloma. Semin Hematol 46:118–126

Rajkumar SV, Fonseca R, Dispenzieri A et al (2000a) Thalidomide in the treatment of relapsed multiple myeloma. Mayo Clinic Proc 75:897–901

Rajkumar SV, Leong T, Roche PC et al (2000b) Prognostic value of bone marrow angiogenesis in multiple myeloma. Clin Cancer Res 6:3111–3116

Rajkumar SV, Hayman S, Gertz MA et al (2002a) Combination therapy with thalidomide plus dexamethasone for newly diagnosed myeloma. J Clin Oncol 20:4319–4323

Rajkumar SV, Gerts MA, Kyle RA et al (2002b) Mayo Clinic Myeloma, Amyloid, and Dysprotidemia Group. Current therapy for multiple myeloma. Mayo Clin Proc 77:813–822

Rajkumar SV, Rosinol L, Hussein M et al (2008) Multicenter, randomized, double-blind, placebo-controlled study of thalidomide plus dexametasone compared with dexamethasone as initial therapy for newly diagnosed multiple myeloma. J Clin Oncol 26:2171–2177

Ramakrishnan V, Timm M, Haug JL et al (2010) Sorafenib, a dual Raf kinase/vascular endothelial growth factor receptor inhibitor has significant anti-myeloma activity and synergizes with common anti-myeloma drugs. Oncogene 29:1190–1202

Rasheed W, Bishton M, Johnstone RW et al (2008) Histone deacetylase inhibitors in lymphoma and solid malignancies. Expert Rev Anticancer Ther 8:413–432

Ravandi F, Cortes E, Jones D et al (2010) Phase I/II study of combination therapy with sorafenib, idrarubicin, and cytarabine in younger patients with acute myeloid leukemia. J Clin Oncol 28:1856–1862

Raver KS, Dixon SC, Figg WD (1998) Inhibition of angiogenesis by thalidomide requires metabolic activation, which is species dependent. Biochem Pharmacol 55:1827–1834

Raza A, Meyer P, Dutt D et al (2001) Thalidomide produces transfusion independence in long-standing refractory anemia of patients with myelodysplastic syndormes. Blood 98:958–965

Redondo-Munoz J, Ugarte-Berzal E, Garcia-Marco JA et al (2008) Alpha 4 beta 1 integrin and 190-kDa Cd 44v constitute a cell surface docking complex for gelatinase B/MMP-9 in chronic leukemic but not in normal cells. Blood 112:169–178

Reikvam H, Hatfield KJ, Lassalle P et al (2010a) Targeting the angiopotein (Ang)/Tie-2 pathway in the crosstalk between acute myeloid leukaemia and endothelial cells: studies of Tie-2 blocking antibodies, exogenous Ang-2 and inhibition of constitutive agonistic Ang-1 release. Exp Opin Invest Drugs 19:169–183

Reikvam H, Hatfield KJ, Oyan Am et al (2010b) Primary human acute myelogeneous leukemia cells release matrix metalloproteinases and their inhibitors: release profile and pharmacological modulation. Eur J Haematol 84:239–251

Reikvam H, Kittang AO, Melve G et al (2013) Targeted anti-leukemic therapy as disease-stabilizing treatment for acute myeloid leukemia relapse after allogenic stem cell transplantation: will it be possible to combine these strategies with retransplantation or donor lymphocyte infusions? Curr Cancer Drug Targets 13:30–47

Relf M, LeJeune S, Scott PA et al (1997) Expression of the angiogenic factors vascular endothelial cell growth factor, acidic and basic fibroblast growth factor, tumor growth factor beta-1, platelet-derived endothelial cell growth factor, placenta growth factor, and pleiotrophin in human primary breast cancer and its relation to angiogenesis. Cancer Res 57:963–969

Rettig WJ, Erickson HP, Albino AP et al (1994) Induction of human tenascin (neuronectin) by growth factors and cytokines: cell-type-specific signals and signaling pathways. J Cell Science 107: 487–497

Ria R, Vacca A, Russo F et al (2004) A VEGF-dependent autocrine loop mediates proliferation and capillarogenesis in bone marrow endothelial cells of patients with multiple myeloma. Thromb Haemost 92:1428–1435

Ria R, Piccoli C, Cirulli T et al (2008a) Endothelial differentiation of hematopoietic stem cells and progenitor cells from patients with multiple myeloma. Clin Cancer Res 14:1678–1685

Ria R, Cirulli T, Giannini T et al (2008b) Serum levels of angiogenic cytokines decrease after radiotherapy in non-Hodgkin lymphomas. Clin Exp Med 8:141–145

Ria R, Todoerti K, Berardi S et al (2009) Gene expression profiling of bone marrow endothelial cells in patients with multiple myeloma. Clin Cancer Res 15:5369–5378

Ribatti D (2004) The involvement of endothelial progenitor cells in tumor angiogenesis. J Cell Mol Med 8:294–300
Ribatti D (2007) The discovery of endothelial progenitor cells. An historical review. Leuk Res 31:439–444
Ribatti D (2009) The paracrine role of Tie-2 expressing monocytes in tumor angiogenesis. Stem Cell Dev 18:703–706
Ribatti D (2011) Antiangiogenic therapy accelerates tumor metastasis. Leuk Res 35:24–26
Ribatti D (2013) Mast cells and macrophages exert beneficial and detrimental effects on tumor progression and angiogenesis. Immunol Lett 152:83–88
Ribatti D, Vacca A (2005) Therapeutic renaissance of thalidomide in the treatment of haematological maligancies. Leukemia 19:1525–1531
Ribatti D, Vacca A (2008) Overview of angiogenesis during tumor growth. In: Figg WJ, Folkman J (eds) Angiogenesis. An integrative approach from science to medicine. Spinger, pp. 161–168
Ribatti D, Djonov V (2012a) Intussusceptive microvascular growth in tumors. Cancer Lett 316:126–131
Ribatti D, Crivellato E (2012b) Mast cells, angiogenesis, and tumour growth. Biochim Biophys Acta 1822:2–8
Ribatti D, Vacca A, Bertossi M et al (1990) Angiogenesis induced by B-cell non-Hodgkin's lymphomas. Lack of correlation with tumor malignancy and immunologic phenotype. Anticancer Res 10:401–406
Ribatti D, Vacca A, Nico B et al (1996) Angiogenesis spectrum in the stroma of B-cell non-Hodgkin's lymphomas. An immunohistochemical and ultrastructural study. Eur J Haematol 56:45–53
Ribatti D, Nico B, Vacca A et al (1998) Do mast cells help to induce angiogenesis in B-cell non-Hodgkin's lymphomas? Br J Cancer 77:1900–1906
Ribatti D, Vacca A, Nico B et al (1999) Bone marrow angiogenesis and mast cell density increase simultaneously with progression of human multiple myeloma. Br J Cancer 79:451–455
Ribatti D, Vacca A, Marzullo A et al (2000) Angiogenesis and mast cell density with tryptase activity increase simultaneously with pathological progression in B-cell non-Hodgkin's lymphomas. Int J Cancer 85:171–175
Ribatti D, De Falco G, Nico B et al (2003a) In vivo time-course of the angiogenic response induced by multiple myeloma plasma cells in the chick embryo chorioallantoic membrane. J Anat 203:323–328
Ribatti D, Molica S, Vacca A et al (2003b) Tryptase-positive mast cells correlate positively with bone marrow angiogenesis in B-cell chronic lymphocytic leukemia. Leukemia 17:1428–1430
Ribatti D, Nico B, Crivellato E et al (2007) The structure of vascular network of tumors. Cancer Lett 248:18–23
Richardson P, Anderson KC (2004) Immunomodulatory analogs of thalidomide: an emerging new therapy in myeloma. J Clin Oncol 22:3212–3214
Richardson P, Hideshima T, Anderson K (2002a) Thalidomide: emerging role in cancer medicine. Annu Rev Med 53:629–657
Richardson PG, Schlossman RL, Weller E et al (2002b) Immunomodulatory drug CC- 5013 overcomes drug resistance and is well tolerated in patients with relapsed multiple myeloma. Blood 100:3063–3067
Richardson PG, Barlogie B, Berenson J et al (2003) A phase 2 study of bortezomib in relapsed, refractory myeloma. N Engl J Med 348:2609–2617
Richardson P, Schlossman R, Jagannath S et al (2004) Thalidomide for patients with relapsed multiple myeloma after high-dose chemotherapy and stem cell transplantation: results of an open-label multicenter phase 2 study of efficacy, toxicity, and biological activity. Mayo Clin Proc 79:875–882
Richardson PG, Weller E, Jagannath S et al (2009) Multicenter, phase I, dose-escalation trial of lenalidomide plus bortezomib for relapsed and relapsed/refractory multiple myeloma. J Clin Oncol 27(5):713–719

Richardson PG, Weller E, Lonial S et al (2010) Lenalidomide, bortezomib, and dexamethasone combination therapy in patients with newly diagnosed multiple myeloma. Blood 116:679–686

Richon VM, O'Brien JP (2002) Histone deacetylase inhibitors: a new class of potential therapeutic agents for cancer treatment. Clin Cancer Res 8:662–664

Ridell B, Norrby K (2001) Intratumoral microvascular density in malignant lymphomas of B-cell origin. APMIS 109:66–72

Roboz GJ, Giles FJ, List AF et al (2006) Phase 1 study of PTK787/ZK222584, a small molecule tyrosine kinase receptor inhibitor, for the treatment of acute myeloid leukemia and myelodysplastic syndrome. Leukemia 20:952–957

Roccaro AM, Hideshima T, Raje N et al (2006) Bortezomib mediates antiangiogenesis in multiple myeloma via direct and indirect effects on endothelial cells. Cancer Res 66:184–191

Rollig C, Brandts C, Shaid S et al (2012) Survey and analysis and prescription pattern of sorafenib in patients with acute myeloid leukemia. Leuk Lymphoma 53 :1062–1067

Rowland TL, Mc Hugh SM, Deighton J et al (1998) Differential regulation by thalidomide and dexamethasone of cytokine expression in human peripheral blood mononuclear cells. Immunopharmacology 40:11–20

Ruan J, Hyjek E, Kermani P et al (2006) Magnitude of stromal hemangiogenesis correlates with histologic subtype of non-Hodgkin's lymphoma. Clin Cancer Res 12:5622–5631

Ruan J, Hajjar K, Rafii S et al (2009) Angiogenesis and antiangiogenic therapy in non-Hodgkin's lymphoma. Ann Oncol 20:413–424

Salmon SE (1975) Expansion of growth fraction in multiple myeloma with alkylating agents. Blood 45:119–129

Salmon SE, Seligman M (1974) B-cell neoplasia in man. Lancet 2:1230–1233

Salven P, Orpana A, Teerenhovi L et al (2000) Simulatenous elevation in the serum concentrations of the angiogenic growth factors VEGF and bFGF is an independent predictor of poor prognosis in non Hodgkin's lymphoma: a single-institution study of 200 patients. Blood 96:3712–3718

Sampaio EP, Sarno EN, Galilly R et al (1991) Thalidomide selectively inhibits tumor necrosis factor alpha production by stimulated human monocytes. J Exp Med 173:699–703

San Miguel JF, Schlag R, Khuageva NK et al (2008) Bortezomib plus melphalan and prednisone for initial treatment of multiple myeloma. N Engl J Med 359(9):906–917

Sato N, Hattori Y, Wenlin D et al (2002) Elevated level of plasma basic fibroblast growth factor in multiple myeloma correlates with increased disease activity. Jpn J Cancer Res 93: 459–466

Scappaticci FA, Smith R, Pathak A et al (2001) Combination angiostatin and endostatin gene transfer induces synergistic antiangiogenic activity in vitro and antitumor efficacy in leukemia and solid tumors in mice. MolTher 3:186–196

Scavelli C, Di Pietro G, Cirulli T et al (2007) Zoledronic acid affects overangiogenic phenotype of endothelial cells in patients with multiple myeloma. Mol Cancer Ther 6:3256–3262

Scavelli C, Nico B, Cirulli T et al (2008) Vasculogenic mimicry by bone marrow macrophages in patients with multiple myeloma. Oncogene 27:663–674

Schaerer L, Schmid MH, Mueller B et al (2000) Angiogenesis in cutaneous lymphoproliferative disorders: microvessel density discriminates between cutaneous B-cell lymphomas and B-cell pseudolymphomas. Am J Dermatopathol 22:140–143

Schey SA, Fields P, Bartlett JB et al (2004) Phase I study of an immunomodulatory thalidomide analog, CC 4047, in relapsed or refractory multiple myeloma. J Clin Oncol 22:3269–3276

Schneider P, Vasse M, Sbaa-Ketata E et al (2001) The growth of highly proliferative acute lymphoblastic leukemia may be independent of stroma and/or angiogenesis. Leukemia 15:1143–1145

Schneider P, Vasse M, Legrand E et al (2003) Have urinary levels of the angiogenic factors, basic fibroblast growth factor and vascular endothelial growth factor, a prognostic value in childhood acute lymphoblastic leukaemia. Br J Haematol 122:163–164

Schneider P, Vasse M, Corbiere G et al (2007) Endostatin variations in childhood acute lymphoblastoid leukaemia. Comparison with basic fibroblast growth factor and vascular endothelial growth factor. Leuk Res 31:629–638

Sehn LH, MacDonald D, Rubin S et al (2011) Bortezomib added to R-CVP is safe and effective for previously untreated advanced-stage follicular lymphoma: a phase II study by the National Cancer Institute of Canada Clinical Trials Group. J Clin Oncol 29:3396–3401

Seidel C, Børset M, Turesson I et al (1998) Elevated serum concentrations of hepatocyte growth factor in patients with multiple myeloma. The Nordic Myeloma Study Group. Blood 91:806–812

Seto M (2010) Genomic profiles in B cell lymphoma. Int J Hematol 92:238–245

Settles B, Stevenson A, Wilson K et al (2001) Down-regulation of cell adhesion molecule LFA-1 and ICAM-1 after in vitro treatment with the anti-TNF-alpha agent thalidomide. Cell Mol Biol 47:1105–1114

Sezer O, Eucker J, Bauhuis C et al (2000) Bone marrow microvessel density is a prognostic factor for survival in patients with multiple myeloma. Ann Hematol 79: 574–577

Sezer O, Jakob C, Eucker J et al (2001a) Seum levels of the angiogenic cytokines basic fibroblast growth factor (bFGF), vascular endothelial growth factor (VEGF) and hepatocyte growth factor (HGF) in multiple myeloma. Eur J Haematol 66:83–88

Sezer O, Niemoller K, Jakob C et al (2001b) Relationship between bone marrow angiogenesis and plasma cell infiltration and serum beta2-microglobulin levels in patients with multiple myeloma. Ann Hematol 80:598–601

Shanafelt T, Zent C, Byrd J et al (2010) Phase II trials of single-agent anti-VEGF therapy for patients with chronic lymphocytic leukemia. Leuk Lymphoma 51:2222–2229

Shannon EJ, Sandoval F (1995) Thalidomide increases the synthesis of IL-2 in cultures of human mononuclear cells stimulated with concanavalin-A, staphylococcal enterotoxin A, and purified protein derivative. Immunopharmacology 31:109–116

Shapiro VS, Mollenauer MN, Weiss A (2001) Endogenous CD28 expressed on myeloma cells up-regulates interleukin-8 production: implications for multiple myeloma progression. Blood 98:187–193

Shinghal S, Mehta J, Desikan R et al (1999) Antitumor activity of thalidomide in refractory multiple myeloma. N Engl J Med 341:1565–1571

Shipp MA, Ross KN, Tamayo P et al (2002) Diffuse large B-cell lymphoma outcome prediction by gene-expression profiling and supervised machine learning. Nat Med 8:68–74

Smith BD, Levis M, Beran M et al (2004) Single agent CEP-701, a novel FLT3 inhibitor, shows biologic and clinical activity in patients with relapsed or refractory acute myeloid leukemia. Blood 103:3669–3676

Smith SM, Grinblatt D, Johnson JL et al (2008) Thalidomide has limited single-agent activity in relapsed or refractory indolent non-Hodgkin lymphoma: a phase II trial of the Cancer and Leukemia Group B. Br J Haematol 140:313–319

Smolej L, Andrys C, Krejsek J et al (2007) Basic fibroblast growth factor (bFGF) and vascular endothelial growth factor are elevated in peripheral blood plasma of patients with chronic lymphocytic leukemia and decrease after intensive fludarabine treatment. Vnitr Lek 53:1171–1176

Sparano JA, Gray R, Giantonio B et al (2004) Evaluating antiangiogenesis agents in the clinic: the Eastern Cooperative Oncology Group Portfolio of Clinical Trials. Clin Cancer Res 10:1206–1211

Spiekermann K, Faber F, Voswinckel R et al (2002) The protein tyrosine kinase inhibitor SU5614 inhibits VEGF-induced endothelial cell sprouting and induces growth arrest and apoptosis by inhibition of c-kit in AML cells. Exp Hematol 30 :767–773

Standker L, Schrader M, Kanse SM et al (1997) Isolation and characterization of the circulating form of human endostatin. FEBS Lett 420:129–133

Steins MB, Padro T, Bieker R et al (2002) Efficacy and safety of thalidomide in patients with acute myeloid leukemia. Blood 99:834–839

Stewart M, Talks K, Leek R et al (2002) Expression of angiogenic factors and hypoxia inducible factors HIF-1, HIF-2 and CAIX in non Hodgkin's lymphoma. Histopathology 40:253–260

Stone RM, De Angelo DJ, Klimek V et al (2005) Patients with acute myeloid leukemia and activating mutation in FLT3 respond to a small-molecule FLt3 tyrosine kinase inhibitor, PKC 412. Blood 105:54–60

Stone RM, Fischer T, Paquette R et al (2012) Phase IB study of the FLT3 kinase inhibitor midostaurin with chemotherapy in younger newly diagnosed adult patients with acute myeloid leukemia. Leukemia 26:2061–2068

Stopeck AT, Unger JM, Rimsza LM et al (2009) A phase II trial of single agent bevacizumab in patients with relapsed, aggressive non Hodgkin lymphoma: Southwest Oncology Group study S0108. Leuk Lymphoma 50:728–735

Storti P, Donofrio G, Colla S et al (2011) HOXB7 expression by myeloma cells regulates their pro-angiogenic properties in multiple myeloma patients. Leukemia 25:527–537

Streubel B, Chott A, Huber D et al (2004) Lymphoma-specific genetic aberrations in microvascular endothelial cells in B-cell lymphomas. N Engl J Med 351:250–259

Strupp C, Germing U, Aivado M et al (2002a) Thalidomide fro the treatment of patients with myelodysplastic syndromes. Leukemia 16:1–6

Strupp C, Aivado M, Germing U et al (2002b) Angioimmunoblastic lymphadenopathy (AILD) may respond to thalidomide treatment: two case reports. Leuk Lymphoma 43:133–137

Sun L, Tran N, Tang F et al (1998) Synthesis and biological evaluations of 3-subsituted indolin-2-ones: a novel class of tyrosine kinase inhibitors that exhibit selectively toward particular receptor tyrosine kinases. J Med Chem 41:2588–2563

Sutherland RM, Mc Credie JA, Inch WR (1971) Growth of multi cell spheroids in tissue culture as a model of nodular carcinomas. J. Natl. Cancer Inst 46:13–17

Tai YT, Podar K, Gupta D et al (2002) CD40 activation induces p53-dependent vascular endothelial growth factor secretion in human multiple myeloma cells. Blood 99: 1419–1427

Takeuchi H, Matsuda K, Kitai R et al (2007) Angiogenesis in primary central nervous system lymphoma (PCNSL). J Neurooncol 84:141–145

Tannock IF (1968) The relation between cell proliferation and the vascular system in a transplanted mouse mammary tumor. Brit J Cancer 22:258–273

Taskinen M, Karjalainen-Lindsberg ML, Nyman H et al (2007) A high tumor-associated macrophage content predicts favourable outcome in follicular lymphoma patients treated with rituximab and cycolophosphamide-doxorubicin-vincristine-prednisone. Clin Cancer Res 13:5784–5789

Thomas DA, Estey E, Giles FJ et al (2003) Single agent thalidomide in patients with relapsed or refractory acute myeloid leukaemia. Br J Haematol 123:436–441

Till KJ, Spiller DG, Harris RJ et al (2005) CLL, but not normal B cells are dependent on autocrine VEGF and alpha4-beta1 integrin for chemokine-induced motility and through endothelium. Blood 105:4813–4819

Trudel S, Ely S, Farooqi Y et al (2004) Inhibition of fibroblast growth factor receptor 3 induces differentiation and apoptosis in t(4;14) myeloma. Blood 103:3521–3528

Tu Y, Gardner A, Lichtenstein A (2000) The phosphatidylinositol 3-kinase/AKT kinase pathway in multiple myeloma plasma cells: roles in cytokine-dependent survival and proliferative responses. Cancer Res 60:6763–6770

Urashima M, Ogata A, Chauhan D et al (1996) Transforming growth factor-beta1: differential effects on multiple myeloma versus normal B cells. Blood 87:1928–1938

Vacca A, Ribatti D, Roncali L et al (1994) Bone marrow angiogenesis and progression in multiple myeloma. Br J Haematol 87:503–508

Vacca A, Ribatti D, Fanelli M et al (1996) Expression of tenascin is related to histologic malignancy and angiogenesis in B-cell non-Hodgkin's lymphomas. Leuk Lymphoma 22: 473–481

Vacca A, Moretti S, Ribatti D et al (1997) Progression of mycosis fungoides is associated with changes in angiogenesis and expression of the matrix metalloproteinase 2 and 9. Eur J Cancer 33:1685–1992

Vacca A, Ribatti D, Iurlaro M et al (1998) Human lymphoblastoid cells produce extracellular matrix-degrading enzymes and induce endothelial cell proliferation, migration, morphogenesis and angiogenesis. Int J Clin Lab Res 28:55–68

Vacca A, Ribatti D, Presta A et al (1999a) Bone marrow neovascularization, plasma cell angiogenic potential and matrix metalloproteinase-2 secretion parallel progression of human multiple myeloma. Blood 93:3064–3073

Vacca A, Iurlaro M, Ribatti D et al (1999b) Antiangiogenesisis produced by nontoxic doses of vinblastine. Blood 94:4143–4155

Vacca A, Ribatti D, Ruco L et al (1999c) Angiogenesis extent and macrophage density increase simultaneously with pathological progression in B-cell non-Hodgkin's lymphoma. Br J Cancer 79:965–970

Vacca A, Ribatti D, Ria R et al (2000) Proteolytic activity of human lymphoid tumor cells. Correlation with tumor progression. Dev Immunol 7:77–88

Vacca A, Ribatti D, Roccaro AM et al (2001 a) Bone marrow angiogenesis in patients with multiple myeloma. Semin Oncol 28: 543–550

Vacca A, Ria R, Presta M et al (2001b) aVb3 integrin engagement modulates cell adhesion, proliferation, and protease secretion in human lymphoid tumor cells. Exp Hematol 29:993–1003

Vacca A, Ria R, Ribatti D et al (2003a) A paracrine loop in the vascular endothelial growth factor pathway triggers tumor angiogenesis and growth in multiple myeloma. Haematologica 88:176–185

Vacca A, Ria R, Semeraro F et al (2003b) Endothelial cells in the bone marrow of patients with multiple myeloma. Blood 102:3340–3348

Vacca A, Scavelli C, Montefusco V et al (2005) Thalidomide downregulates angiogenic genes in bone marrow endothelial cells of patients with active multiple myeloma. J Clin Oncol 23:5334–5346

Veiga JP, Costa F, Sallan SE et al (2006) Leukemia-stimulated bone marrow endothelium promotes leukemia cell survival. Exp Hematol 35:610–621

Verheul HM, Panigrahy D, Yuan J et al (1999) Combination oral antiangiogenic therapy with thalidomide and sulindac inhibits tumour growth in rabbits. Br J Cancer 79:114–118

Verstovsek S, Kantarjian H, Manshouri T et al (2002) Prognostic significance of tie-1 protein expression in patients with early chronic phase in chronic myeloid leukemia. Cancer 94:1517–1521

Vincent L, Jin DK, Karajannis M et al (2005) Fetal Stromal-Dependent Paracrine and Intracrine Vascular Endothelial Growth Factor-A/Vascular Endothelial Growth Factor Receptor-1 Signaling Promotes Proliferation and Motility of Human Primary Myeloma Cells. Cancer Res 65:3185–3192

Wakabayashi M, Miwa H, Shikami M et al (2004) Autocrine pathway of angiopoetins-Tie-2 system in AML cells. The Hematol J 5:353–360

Wang ES, Teruya-Feldstein J, Wu Y et al (2004) Targeting autocrine and paracrine VEGF receptor pathways inhibits human lymphoma xenografts in vivo. Blood 104:2893–2902

Weber D, Rankin K, Gavino M et al (2003) Thalidomide alone or with dexamethasone for previously untreated multiple myeloma. J Clin Oncol 21:16–19

Wedge SR, Ogilvie DJ, Dukes M et al (2002) ZD6474 inhibits vascular endothelial growth factor signaling, angiogenesis, and tumor growth following oral administration. Cancer Res 62:4645–4655

Wedge SR, Kendrew J, Hennequin LF et al (2005) AZD2171: a highly potent, orally bioavailable, vascular endothelial growth factor receptor-2 tyrosine kinase inhibitor for the treatment of cancer. Cancer Res 65:4389–4400

Wegiel B, Ekberg J, Talasila KM et al (2009) The role of VEGF and a functional link between VEGF and p27Kip1 in acute myeloid leukemia. Leukemia 23:251–261

Wellmann S, Guschmann M, Griethe W et al (2004) Activation of the HIF pathway in childhood ALL, prognostic implications of VEGF. Leukemia 18: 926–933

Wiernik PH, Lossos IS, Tuscano JM et al (2008) Lenalidomide monotherapy in relapsed or refractory aggressive non-Hodgkin's lymphoma. J Clin Oncol 26:1–6

Williams S, Pettaway C, Song R et al (2003) Differential effects of the proteasome inhibitor bortezomib on apoptosis and angiogenesis in human prostate tumor xenografts. Mol Cancer Ther 2(9):835–843

Wilson EA, Jobanputra S, Jackson R et al (2002) Response to thalidomide in chemotherapy-resistant mantle cell lymphoma: a case report. Br J Haematol 119:128–130

Witzig TE, Wiernik PH, Moore T et al (2009) Lenalidomide oral monotherapy produces durable responses in relapsed or refractory indolent non-Hodgkin's lymphoma. J Clin Oncol 27:5404–5409

Wolf JE, Hubler WR (1975) Tumour angiogenic factor associated with subcutaneous lymphoma. Br J Dermatol 92:273–277

Wood J, Bonjean K, Ruetz S et al (2002) Novel antiangiogenic effects of the bisphosphonate compound zoledronic acid. J Pharmacol Exp Ther 302:1055–1061

Wuillème-Toumi S, Robillard N, Gomez P et al (2005) Mcl-1 is overexpressed in multiple myeloma and associated with relapse and shorter survival. Leukemia 19:1248–1252

Yee KW, O'Farrell AM, Smolich BD et al (2002) SU5416 and SU5614 inhibit kinase activity of wild-type and mutant FLT3 receptor tyrosine kinase. Blood 100:2941–2949

Yetgin S, Yenicesu I, Cetin M et al (2001) Clinical importance of serum vascular endothelial and basic fibroblast growth factors in children with acute lymphoblastic leukemia. Leuk Lymphoma 42:83–88

Zangari M, Anaissie E, Barlogie B et al (2001) Increased risk of deep-vein thrombosis in patients with multiple myeloma receiving thalidomide and chemotherapy. Blood 98:1614–1615

Zangari M, Anaissie E, Stopeck A et al (2004) Phase II study of SU5416, a small molecule vascular endothelial growth factor tyrosine kinase receptor inhibitor, in patients with refractory multiple myeloma. Clin Cancer Res 10:88–95

Zahiragic L, Schliemann C, Bieker R et al (2007) Bevacizumab reduces VEGF expression in patients with relapsed and refractory acute myeloid leukemia without clinical antileukemic activity. Leukemia 21:1310–1312

Zhang B, Gojo I, Fenton RG (2002) Myeloid cell factor-1 is a critical survival factor for multiple myeloma. Blood 99:1885–1893

Zhang H, Vakil V, Braunstein M et al (2005) Circulating endothelial progenitor cells in multiple myeloma: Implications and significance. Blood 105:3286–3294

Zhao WL, Mourah S, Mounier N et al (2004) Vascular endothelial growth factor-A is expressed both on lymphoma cells and endothelial cells in angiommunoblastic T-cell lymphoma and related to lymphoma progression. Lab Invest 84:1512–1519

Index

D. Ribatti, *Angiogenesis and Anti-Angiogenesis in Hematological Malignancies*,
DOI 10.1007/978-94-017-8035-3, © Springer Science+Business Media Dordrecht 2014

Zeitfracht Medien GmbH
Ferdinand-Jühlke-Straße 7
99095 Erfurt, Deutschland
produktsicherheit@kolibri360.de